INSECTS OF
STORED
PRODUCTS

INSECTS OF
STORED
PRODUCTS

David Rees

With photographs by Vanna Rangsi

CSIRO
PUBLISHING

MANSON
PUBLISHING

National Library of Australia Cataloguing-in-Publication entry
Rees, David.
Insects of stored products.

   Bibliography.
   Includes index.
   ISBN 0 643 06903 8.
   1. Insect pests. 2. Food storage pests. I. Title.

   632.7

Published exclusively in Australia, New Zealand and the Americas, and non-exclusively in other territories of the world (excluding Europe, Africa, and the Middle East), by:

CSIRO PUBLISHING
150 Oxford Street (PO Box 1139)
Collingwood VIC 3066
Australia

Telephone:	+61 3 9662 7666
Local call:	1300 788 000 (Australia only)
Fax:	+61 3 9662 7555
Email:	publishing.sales@csiro.au
Web site:	www.publish.csiro.au

Published exclusively in Europe, Africa and the Middle East, and non-exclusively in other territories of the world (excluding Australia, New Zealand and the Americas) by **Manson Publishing Ltd**, with the **ISBN 1-84076-060-5.**

A CIP catalogue record for this book is available from the British Library.

For full details of all Manson Publishing Ltd titles please write to:
Manson Publishing Ltd
73 Corringham Road
London NW11 7DL, UK

Telephone:	+44(0)20 8905 5150
Fax:	+44(0)20 8201 9233
Website:	www.manson-publishing.com

Front cover: Main photo by David Rees. Strip photos (from left) by Vanna Rangsi (first three images), John Green and David Rees, all CSIRO

Back cover: (clockwise from top left) by Yonglin Ren, David McClenaghan, David Rees, David McClenaghan, David Rees and Vanna Rangsi, all CSIRO

Set in Minion 9.5/11
Cover and text design by James Kelly
Printed in Australia by Impact Printing

Table of contents

Acknowledgements

First I would like to thank the Head of the Stored Grain Research Laboratory (SGRL), Mr Colin Waterford and his predecessors Dr Jonathan Banks and Dr Jane Wright for their interest, encouragement, permission and access to resources to allow me to undertake this project. In addition, I thank Dr Bruce Halliday and Jan van S. Graver for their advice and review of the manuscript as it evolved. I also acknowledge the assistance of unknown external reviewers and editors.

I thank Ann Crabb and the staff of **CSIRO** PUBLISHING for taking this project on, for keeping me going and for guiding this project to completion.

I also thank my parents for their encouragement in keeping me going on the project through a difficult time in my life.

Assembling the illustrations which are such a central part of this book would not have been possible without the help of colleagues at CSIRO and overseas. Of CSIRO Entomology staff I thank Dr John LaSalle, Dr Andrew Calder, Dr Rolf Oberprieler, Mr Tom Weir and Dr Marianne Horak for allowing me to use specimens from the Australian National Insect Collection. In particular, I thank Vanna Rangsi for her considerable efforts in taking and editing the majority of the images used in this book. Images were also taken by John Green, David McClenaghan and Noel Starick. Assistance was also provided by Anne Hastings, Chris Hunt and Soussanith Nokham. I also thank my technical assistants, Debbie Parsons, Bradley Brown, and the late Nina Patelis, for maintenance and supply of live insect cultures.

In addition, I thank Mr Justin Dixon, Invertebrate Supply Officer, Central Science Laboratory, York, UK, for specimens of *Pseudeurostus hilleri*, *Tipnus unicolor*, *Dermestes haemorrhoidalis*, *Coelopalorus foveicollis*, *Tribolium audax*, *Tribolium destructor* and *Tribolium madens*. Also I thank Dr Rick Hodges of the Natural Resources Institute, Chatham, UK, for the supply and permission to use figures 27 and 106 and Dr Michael Toews, Research Entomologist, United States Department of Agriculture, Grain Marketing and Production Research Center, Manhattan, Kansas, USA, for specimens of *Cynaeus angustus*.

Finally, this book is in memory and respect to previous generations of workers in the discipline of stored products entomology, whose efforts give us the foundations of knowledge that we rely on today and form the basis of the information presented here.

Preface

Central to the survival and prosperity of complex human civilisations is their ability to store durable foodstuffs such as grains between harvests to ensure a constant food supply. Even when stored under good physical conditions, durable foodstuffs are at serious potential risk from destruction and spoilage by insects, especially in warm and tropical regions. Insects are small and difficult to detect at initial infestation and can easily enter all but the most well-sealed storage structures. Under optimal conditions insect populations develop very quickly and can cause severe damage both in terms of loss of quantity and quality. Until the widespread adoption of fumigants and grain protectants, control and elimination of the pests was very difficult and heavy losses were the norm. Indeed, in many societies, the 'sudden' appearance of heavy infestations in stored commodities were, and sometimes still are, regarded as 'inherent vice' – an inevitable consequence of storing grain or part of the grain itself.

The ability to identify the different species of insect directly associated with stored products, together with access to information about their biology and pest status, is important to be able to mount effective control measures against them. Much of this information is currently scattered through specialist scientific literature which is often hard to access. The aim of this publication is provide a convenient entry point to information on the field identification, economic importance, pest status, distribution, life history and ecology of the most frequently encountered insects found in grain and other stored durable products. References are provided to the scientific literature for further information and study.

Insects as pests of stored products

Introduction and scope

In terms of the number of known species, insects are the most diverse form of life on earth. The exact number of species will probably never be known but is likely to exceed one million species. Insects are currently divided into 32 orders. Only three orders: beetles (Coleoptera), moths (Lepidoptera) and psocids or booklice (Psocoptera) contain species that are major pests of stored products. Bugs (Hemiptera) and wasps (Hymenoptera) are also found in stored products as predators and parasitoids of the pest species mentioned above. Members of other orders can also be present as incidental scavengers such as silverfish (Thysanura), cockroaches (Blattodea) and flies (Diptera). Termites (Isoptera) can also cause incidental damage to stored products as they feed on wooden structures.

This book covers the major species of beetles, moths, psocids, parasitic wasps and predatory bugs that infest dry durable food, fibre and skin products worldwide. Such materials include seeds (cereal grain, pulses, oilseeds, nuts, beverage crops), dried tubers, dried fruit, herbs and spices, dried fish and meat, museum and herbarium artefacts and specimens, hides, skins and wool. Many of these materials are major items of trade and as such have considerable economic value.

Wood-boring insects are not included in this publication except where they are also a pest of one or more of the commodities listed above. Incidental pests of the built environment such as cockroaches, silverfish and flies are also not covered.

In addition to insects, many mite species also infest durable stored products. These are not covered in this work. Readers wishing to know more about mites in stored products should refer to Hughes (1976).

Origins of stored product insects

Between 5000 and 10 000 years ago human society commenced settled agriculture and began to produce and store large quantities of dried organic materials such as grains, fibres and skins. A vast new resource then became available which attracted a select band of insects that feed on dry material of animal and plant origin. These insects came from a variety of natural habitats, which include:

- Under bark of trees: **Coleoptera** – Cleridae, Laemophloeidae, Ptininae, Tenebrionidae, Silvanidae, Trogossitidae; **Psocoptera**
- Seeds: **Coleoptera** – Curculionidae, Bruchinae; **Lepidoptera** – Gelechiidae
- Leaf litter: **Lepidoptera** – Oecophoridae, Pyralidae; **Psocoptera**
- Dead and ripening fruit: **Coleoptera** – Nitidulidae; **Lepidoptera** – Pyralidae
- Wood, shoot and tuber borers: **Colcoptera** – Anobiidae, Anthribidae, Bostrichidae
- Fungi and mould: **Coleoptera** – Latridiidae, Mycetophagidae; **Psocoptera**
- Carrion and dead animals: **Coleoptera** – Dermestidae, Cleridae
- Nests of wasps, birds and mammals: **Coleoptera** – Dermestidae, Ptininae, Tenebrionidae; **Lepidoptera** – Tineidae.

Many storage pests have been associated with human activity for a long time. *Tribolium confusum* (Coleoptera: Tenebrionidae) *Oryzaephilus surinamensis* (Coleoptera: Silvanidae) and *Sitophilus granarius* (Coleoptera: Curculionidae) were well established as pests of grain in ancient Egypt.

Originally, many species had restricted distributions, but thousands of years of trade has resulted in most pest species now having global distributions. The process of spread is still continuing, as witnessed by the colonisation of sub-Saharan Africa during the 1980–1990s by the tropical American pest *Prostephanus truncatus* (Coleoptera: Bostrichidae).

Impacts of infestation

Infestation of stored products by insects results in a variety of damage and economic loss including:

- physical loss of commodity – by direct consumption
- spoilage and loss of commodity quality – down grading due to physical and nutritional damage
- waste of the effort taken in growing, handling, manufacturing and storing commodities which are destroyed by insect infestation
- encouragement of mould growth – including of those fungi that produce mycotoxins
- contamination of commodities with insect bodies, waste products etc. – some of which are toxic, repulsive or allergenic
- rejection by consumers (both human and animal) of infested commodities and the resultant social and legal costs
- costs associated with application of measures to control and prevent infestations
- risks to health, safety and the environment relating to use of pesticides and fumigants
- restriction of trade and damage to economies and the environment caused by inadvertent introduction of 'quarantine' pest species.

In stable, well organised societies, infestations of storage pests are mostly held under control. However, the costs of keeping insects under control are significant and damaging infestations can occur if mistakes or neglect occur. Over the life of a given batch of commodity, protection against insect attack may account for several percent of its value. This cost mounts up over the years and over millions of tonnes of commodities handled. Businesses and consumers generally have a zero tolerance to infestation of products destined for direct human consumption and as a result the economic loss resulting from the simple presence of insects in such products can be considerable. A box of chocolates containing a single moth larva is worse than worthless. It may result in a fine from the environmental health department, negative publicity and lost consumer trust in a brand and/or legal action.

Pests thrive under turmoil. Significant physical damage of food by storage pests is most likely to occur in societies least able to cope, such as those under stress by poverty and additionally by famine, new pests, natural disasters and war. The need to protect food stocks during World Wars I and II was a major initial stimulus into research on stored product pests and means to control them. Much of the basic scientific data collected at this time forms the foundation of commodity protection today.

Societies in tropical regions which rely upon subsistence agriculture are especially vulnerable to losses caused by storage insects. Annual weight loss due to storage pests may range between 2–9% under normal circumstances. The impact on physical and nutritional quality is less understood but is likely to be greater. The ability of such communities to replace grain 'stolen' by pests may be very limited and as a result, badly affected communities go hungry. Nationally, the impact can be considerable; replacing the tens or hundreds of thousands of tonnes or more of grain lost to pests costs millions of dollars annually.

International trade and the success of farming enterprises can be badly affected by the arrival or threatened arrival of new pests. The costs associated with maintenance of effective quarantine is considerable but are dwarfed by the costs incurred when things go wrong. Since the early 1980s, the poorest farmers of sub-Saharan Africa have had to cope with the relentless spread of the larger grain borer *Prostephanus truncatus*, a pest inadvertently introduced from the Americas. Important grain exporters such as the USA and Australia take considerable trouble to prevent the arrival and establishment of the Khapra beetle, *Trogoderma granarium* (Coleoptera: Dermestidae). Establishment of this insect could cause considerable disruption to grain export trade from these countries.

The impact on human and animal health of allergenic substances present as a result of insect infestation is not well understood but may be significant. What is better understood is the potential impact on health of fungal toxins resulting from mould growth that is encouraged by insect infestation as well as by poor storage and drying practices. This impact may be considerable in parts of the humid tropics where insect and mould development is especially rapid. In many parts of the world there is an increasing desire by consumers not to consume food that has been treated with pest control chemicals. Registration and use of these chemicals is being increasingly strictly regulated to minimise risk to users, consumers and the environment. However, the debate on their use needs to include an understanding of what other dangers use of pesticides and fumigants protects consumers from, for example the risk of hunger, loss of food quality and security, and the risk of chronic poisoning by natural fungal toxins.

Use of chemicals to control storage insects can have considerable environmental consequences. Methyl bromide, a fumigant widely used to control such pests, especially on commodities in international trade, has been found to be a potent atmospheric ozone depletor. Without the ozone layer protecting us from harmful solar radiation, life on earth would become increasingly difficult. As a result, methyl bromide is being withdrawn by international treaty and the search is on for replacements.

Feeding strategies

Insects infesting stored products feed and live in a number of ways which include:

- Commodity feeders (primary and secondary pests)
- Fungal feeders
- Predators
- Parasitoids
- Scavengers
- Foragers and accidentals.

Commodity feeders

Insects that feed directly on a commodity, especially seeds and products made from them, are often divided into primary pests – insects that can attack intact seeds and secondary pests that require the commodity to be damaged before they attack. In reality the situation is more complex. Each pest species requires its own level of 'damage' before it is able to successfully breed on a commodity. At one extreme are insects which clearly fill the role of primary pests as they are able to attack undamaged seed. Examples include bruchids, bostrichids, weevils and the moth *Sitotroga cerealella*. In reality, grain never enters storage totally undamaged. Grain accumulates damage such as chipped seed coats as a result of harvesting, handling, transporting, cleaning and drying. Such damage increasingly allows attack by secondary pests such as *Tribolium* spp., *Oryzaephilus* spp., *Cryptolestes* spp. and psocids *Liposcelis* spp. Damage previously caused by pre-harvest pests and by primary storage pests will also assist secondary pests.

In milled products such as flour, secondary pests dominate. Flour is after all highly 'damaged' grain. Indeed the primary pests listed above are unable to attack milled products unless they are highly compacted or are processed into a solid form such as pasta or milled rice. Secondary pests are often selective as to which parts of the commodity they attack – many preferentially feed on the germ of grains. Secondary pests form the bulk of the pests attacking complex processed and manufactured food products such as breakfast cereals, chocolate and compound animal foods. Many of these pests are highly flattened in form and are able to easily enter packaged goods.

Primary pests tend to have a more restricted host range than secondary pests. Many secondary pests, such as *Trogoderma* spp. (Coleoptera: Dermestidae) and *Tribolium* spp. attack a very wide range of materials of both animal and plant origin.

Some commodities, e.g. copra and dried fish, inherently provide opportunities for access by insects, as the processes used to make them always produces cracks and crevices. Here the classification into primary and secondary pests is not very meaningful.

Which of the primary or secondary pests species present is the most important depends on the situation. For subsistence maize producers in tropical Africa, primary pests such as *Sitophilus zeamais* (Coleoptera: Curculionidae) and *P. truncatus* are of greatest concern. For a manufacturer of chocolates a secondary pest *Plodia interpunctella* (Lepidoptera: Pyralidae) is likely to be the major problem.

Commodities vary in their susceptibility to attack. Commodities that contain toxins and antifeedant chemicals, tend to have fewer and more specific pests, like the bruchid beetles that have evolved to attack pulses. Even dried tobacco – a material which contains the insecticide nicotine, has a characteristic suite of pests – notably *Lasioderma serricorne* (Coleoptera: Anobiidae) and *Ephestia elutella* (Lepidoptera: Pyralidae). In comparison, cereal grains and their products are attacked by a wide range of pest species.

Fungal feeders

Many storage pests supplement their diet by feeding on mould and mould spores. This provides additional nutrients that are absent or unavailable directly from the commodity itself. Other species, for example beetles of the families Latridiidae and Mycetophagidae, are mostly obligatory mould feeders and cannot survive on clean dry grain. Fungal feeders are often present on ripening grain and usually die out in storage but may continue to breed in poorly stored grain or in grain heavily infested with other insects.

Predators

Many storage pests, for example beetles of the families Cleridae, Tenebrionidae and Trogossitidae, will also prey on other insects present including members of their own species. Hemiptera and members of the beetles of the family Histeridae are obligate predators. One histerid, *Teretrius nigrescens*, has been deliberately introduced in Africa as a bio-control agent to prey on the larger grain borer *Prostephanus truncatus*.

Parasitoids

Beetle and moth pests of stored products may be attacked by a number of parasitic wasps. These wasps lay their eggs on or in the eggs or larvae of their host. Wasp larvae then develop on host tissue, eventually killing their host as they emerge as mature larvae prior to pupation or as an adult wasp depending on species. Presence of large numbers of these wasps in a store usually indicates established pest infestations. There has been interest in using these wasps as bio-control agents, especially against structural infestations in premises that process organic grade produce.

Scavengers

A number of species feed on dead insects and other dried material of animal origin. These include members of the families Ptininae, Cleridae and Dermestidae. Many of these insects are also important pests of stored products of animal origin such as wool, hides, skins and dried fish.

Foragers and accidentals

A range of insects associated with the built environment are often found in food stores. These include ants and cockroaches as foragers and wood-boring beetles and termites which attack storage structures.

In addition to the species covered above, many other species can accidentally contaminate stored products. These often include general predators such as carabid beetles. In some areas and in some seasons problems can occur with field insects being harvested along with the crop. While these generally do not survive long in the stored commodity their presence may cause customer rejection, taint and contamination and, when dead, can become a source of food for stored product pests.

Impact of environmental conditions on population growth

Temperature, relative humidity and commodity moisture content profoundly affect the rate at which stored product pests can multiply and hence become pests.

Temperature

Biochemical processes that are the core of life are highly influenced by temperature. Unlike mammals and birds, insects are unable to maintain a constant body temperature and as a result their body temperature rises and falls with the temperature of their surroundings. As a result, the ability of an insect to breed or survive is highly dependent upon the temperature of its surroundings.

Stored product pests breed within a temperature range of about 15–42°C. No species covers this range completely and each varies in its tolerance to either the high or low extremes of this range. Most important pest species breed within a range of 18–38°C. Above the minimum temperature, the rate of population growth increases with temperature until an optimum is reached, which typically falls between 25–33°C. Beyond this point the rate of population growth drops off quickly as heat stress related factors become more important.

Typically most species breed at least 10 times faster at their optimum temperature than at their minimum, with a resultant difference in pest status. Storage pests are much more serious where mean temperatures are close to optimal values, such as in the humid tropics. Grain and other bulk-stored commodities hold their temperature well, especially when the mass involved is large. Even in temperate regions where grain is harvested warm in summer, insects can still be a problem well into the winter months as grain can still retain its warmth.

Insects can survive periods with temperatures above or below those suitable for breeding. As temperatures drop, insect activity declines to nothing. Eventually they will die but this may take a long time – even months or years. Some species are tolerant enough to survive moderate freezing, especially if the onset of such conditions is gradual. Strains of widely distributed species, for example *Cryptolestes* spp. (Coleoptera: Laemophloeidae), originating in seasonally cold areas such as Canada are more likely to be able to do this than tropical strains of the same species. Rapid exposure to temperatures of −20°C will kill most insects in a few days.

Upon exposure to extreme conditions and/or the lack of food larvae of some species can enter a state of 'suspended animation' known as diapause. In this state, biological activity is minimised and tolerance to extreme conditions is maximised. Some, such as larvae of *Trogoderma* species (Coleoptera: Dermestidae), can remain in this state for five years or more, waiting for conditions to improve.

As temperatures increase above those at which they can breed, insects become more and more affected by heat stress. Above 50°C the time taken to die is further shortened as such conditions effectively begin to cook the insect. Exposure to temperatures above 60°C results in death in seconds.

Relative humidity and commodity moisture content

Relative to most insects, species which attack dried commodities are highly resistant to moisture loss. However, most species breed fastest under humid conditions, typically 60–80%, which in terms of grain moisture content is roughly equivalent to 13–15%. At higher humidities, mould growth can become a problem and this may hinder the development of some species. The impact on population development of humidities below 60% varies considerably between species. Generally the drier it is the greater the mortality, in particular of early life stages. Some important grain storage pests are very tolerant of low humidities – notably the beetles *Cryptolestes ferrugineus*, *Tribolium castaneum* and *Rhyzopertha dominica* and the moth *Cadra cautella* (Lepidoptera: Pyralidae).

Interaction between response to relative humidity and temperature

Relative humidity and temperature affect different species in different ways. At optimal conditions, pest species that dominate tend to be those that can breed most rapidly on the commodity concerned and/or prey on other potential competitors. At more extreme conditions the species that dominate are those that are best able to survive and breed, even if they only breed slowly. Such species may often be out-competed under more favourable conditions. The table below illustrates possible changes in pest communities that could occur in wheat or barley when stored in a wide range of environments. Some species such as *Trogoderma granarium* tend to be associated with hot dry conditions because relative to their competitors they do well under these conditions. Ptinid beetles, for example, breed slowly and are usually minor pests; however, they can breed in much colder conditions than other pests and so then they become relatively important.

Impact of temperature and relative humidity on likely pest communities found in stored wheat or barley

Relative humidity (r.h.)	Low temperature < 25° Temperate regions	Medium temperature 25–35°C Warm temperate to tropical	High temperature > 35°C Hot tropics
Low < 50%	*Cryptolestes capensis* *Tribolium confusum*	*Cryptolestes ferrugineus* *Oryzaephilus surinamensis* *Tribolium castaneum* *Rhyzopertha dominica*	*Cryptolestes ferrugineus* *Tribolium castaneum* *Trogoderma granarium*
Medium 50–75%	*Cryptolestes capensis* *Sitophilus granarius* *Tribolium confusum* *Oryzaephilus surinamensis*	*Cryptolestes ferrugineus* *Oryzaephilus surinamensis* *Rhyzopertha dominica* *Sitophilus oryzae* *Tribolium castaneum*	*Cryptolestes ferrugineus* *Rhyzopertha dominica* *Tribolium castaneum* *Trogoderma granarium*
High > 75%	*Oryzaephilus surinamensis* *Ptinus* spp. *Sitophilus granarius* *Tribolium confusum*	*Cryptolestes ferrugineus* *Oryzaephilus surinamensis* *Rhyzopertha dominica* *Sitophilus oryzae* *Tribolium castaneum*	*Cryptolestes ferrugineus* *Liposcelis paeta* *Latheticus oryzae* *Rhyzopertha dominica* *Tribolium castaneum*

How to use this guide

Information on insect species is presented in 23 sections. In the case of the beetles and moths, each section covers a family. Some sections are further divided into subsections covering a significant genus or species within that family. The remaining three orders of insects (psocids, bugs and wasps) are each covered by a single section.

Information within each section or subsection is laid out in a similar way, which is detailed below.

Classification and names

Common and scientific names are given for the family, genus or species in question. In some sections only a selection of the more important genera or species as pests of stored products are covered and this is noted by the annotation 'selected genera' or 'selected species'. Below the section heading a list of the more important pest species or genera of that family or group is given. Common names are given for species where such names are in widespread usage.

Summary

A summary table lists important points concerning feeding strategies, economic importance, distribution, life history and biology.

Introduction

A brief description is given of the lifestyle of the order, family, genus or species in question.

Identification

A description is given of the species or genera and the major features used for identification. As a field guide, this book covers the identification process to the extent that is easily possible in the field with a binocular microscope. References are made to relevant specialised keys should identifications of greater precision be required.

Images of insects (Figures 1 to 228) are annotated with scale bars and arrows. Unless otherwise indicated, the scale bar represents 1 mm. Arrows on images point to important charateristics useful in identifiation.

Life cycle

Details are given on the process of development from egg to adult with particular reference to the life style of larvae/nymph and adult stages.

Physical limits and optimum rate of multiplication

Where recorded in the scientific literature, the range and optimal conditions of temperature and humidity under which populations of the species in question will grow is given. In addition, an indication is given of the maximum speed by which a population of the insect can increase in a month. Information presented in this section is based mainly on data published in Gorham (1991), Howe (1965) and White (1995).

The relevance of this laboratory data to real life situations will vary. Information presented should be treated as a guide. Optimal conditions and temperature and humidity limits are likely to be a reasonably accurate reflection of what may occur in real world conditions. However, rates of population growth achieved will be highly variable and will depend to a large extent on the presentation and nutritional quality of the commodity infested.

Economic importance

Details are given on pest status, what commodities are attacked and under what circumstances this occurs. Pest status is highly affected by climate and nature of commodity infested. In addition, the simple presence of any insect, even if only a minor pest, can be important from an economic or quarantine perspective.

Type of damage and symptoms

A description of the nature of the damage inflicted by the pest or group of pests – especially if the damage is distinctive and is an aid to the identity of the pest in question.

Ecology

Information is given about interactions with other species and the storage environment, also about the occurrence of the species in question in natural, non-storage habitats.

Monitoring

Information is given on methods to best detect the insect concerned, with particular reference to commercially available trapping systems and simple easy-to-make traps which the reader can construct for themselves.

Geographical distribution

Comments on distribution are given. In most cases, important pest species now have a world wide or a pan-tropical distribution. Most species are more often encountered in tropical rather than temperate regions. However, many tropical species not established in temperate areas are frequently intercepted there on infested imports. Summary tables showing distribution based on bio-geographic zones are given in most datasheets. Zones are based on those used by Aitken (1975) and are shown on the map below. Distribution data given is based on data published in Aitken (1975, 1984), Bousquet (1990), Carvalho (1979), Haines (1974, 1981), Harney (1993), Lienhard (1990), Peacock (1993), Prakash *et al.* (2003), Semple (1985), Wang Suya and Shan Meijing (1998) and Zakladoi and Ratanova (1987). Data examined and presented may not be comprehensive for all species and it is possible that some species may be found in regions not listed in this book.

In additition, in summary tables an indication of the pest or other status is given. For pest species this is shown as:

• – minor pest, rarely causes much damage

•• – pest capable of causing damage in stored products

••• – pest capable of significant damage

•••• – major pest rapidly capable of devastating damage.

Bio-geographic zones used for presentation of distribution data are given in Figure A.

Figure A. Bio-geographic zones used in this book: 1 USA and Canada, 2 Central and South America (including Mexico), 3 Europe and northern Asia, 4 Mediterranean basin, 5 Africa, 6 South and South-east Asia, 7 Australia and Oceania.

References

References are given herein to specialist publications containing more information about the group of insects concerned, especially those which describe the process of identification to species.

Orders of insects associated with stored products

Orders of insects likely to be encountered in stored products can be identified using Figure B and the key on page 10. Those covered in specific detail in this guide are indicated. A full list of species covered in this guide is given in the Index to species.

Beetles (Coleoptera)
Page 11

Beetles (Coleoptera)
Page 11

Moths (Lepidoptera)
Page 121

Psocids (Psocoptera)
Page 137

Psocids (Psocoptera)
Page 137

Bugs (Hemiptera)
Page 143

Parasitic wasps
(Hymenoptera)
Page 147

Ants and bees
(Hymenoptera)
Not covered

Silverfish (Thysanura)
Not covered

Termites (Isoptera)
Not covered

Cockroaches (Blattodea)
Not covered

Flies (Diptera)
Not covered

Arthropods: spiders
Not covered

Arthropods: scorpions
Not covered

Arthropods: mites,
Not covered

Figure B. Orders of insects and other arthropods associated with stored products.

Key to orders of insects associated with stored products

1   Three pairs of legs attached to thorax, three major body regions (head, thorax, abdomen), antennae present, often with one or two pairs of wings – (Insects) 2
    Four or more pairs of legs, one or two body regions, antennae absent, never with wings
    . . . .**other arthropods – spiders, mites etc. (not covered by this publication)**

2   Mouthparts designed for sucking (Figures H and 223) 3
    Mouthparts designed for chewing (Figure C2) 5

3   One pair of wings (hindwings reduced to small projections) . . . . **Flies (Diptera)** –
    scavengers and incidentals – **not covered by this publication**
    Two pairs of wings present 4

4   Mouthparts stored curled like coiled spring, wings soft and delicate covered with thousands of overlapping tile-like scales . . . . . **Moths (Lepidoptera) – Pages 121–136**
    Mouthpart like needle, when not in use held straight underneath body between bases of legs. Fore wings not covered in scales, base of fore wing often opaque and thickened, hindwing and tip of fore wing are membranous . . . **Bugs (Hemiptera) – Pages 143–145**

5   Hard wing-cases (elytra) instead of fore wings (Figure C1). Elytra meet in straight line down middle of abdomen and never overlap. No obvious appendages protruding from tip of abdomen . . . . **Beetles (Coleoptera) – Pages 11–120**
    Fore wings not as above or wingless. Sometimes abdomen has obvious appendages protruding from tip 6

6   Body constricted at front of abdomen to form waist (Figure 224), often with a long needle-like appendage at the tip of the abdomen (ovipositor) . . . .
    ants, bees and wasps (Hymenoptera) – parasitic wasps – Pages 147–150, ants, scavengers in stores – **not covered by this publication**
    Body not constricted to form waist 7

7   Tip of abdomen without obvious projections 8
    Tip of abdomen with obvious projections 9

8   Small size – most 1 mm, all less than 3 mm long, antennae long and hair-like . . . .
    . . . . **Psocids (Psocoptera) – Pages 137–142**
    Larger size – usually 5 mm or larger, winged or wingless, usually many individuals together, different body forms present, e.g. 'soldiers' with enlarged heads and mouthparts . . . .
    Termites (Isoptera) – usually wingless worker cast as pests of wooden structures – **not covered by this publication**

9   Tip of abdomen with three long filamentous appendages up to half body length, elongate and flattened, silver grey in colour, up to 15 mm in length, wingless . . . .
    . . . . . . **Silverfish (Thysanura)** – scavengers in stores – **not covered by this publication**
    Tip of abdomen with two short appendages, body flattened, most species oval in shape brown or black, when wings are present fore wings are leathery and at rest they overlap down the middle of the abdomen . . . .
    . . . . .**Cockroaches (Blattodea)** – scavengers in stores – **not covered by this publication**

Beetles (Order: Coleoptera)

The Coleoptera contains by far the largest number of known species of any insect order. Adult beetles are characterised by having fore wings modified into hard protective structures known as elytra (see below). Metamorphosis is complete: eggs hatch into larvae, the feeding and growing stage, which then pupate to become pupae, a transition stage from which the adult emerges. Both adults and larvae possess biting mouth parts.

Beetles occur in every environment in which insects can survive and many are important pests of agriculture, horticulture and forestry. Some 45 families of beetles have been recorded from stored products. These range from important pest species through to mould feeders, scavengers and accidentals. They form the majority of important pest species of stored products – in particular members of the families Anobiidae, Bostrichidae, Chrysomelidae, Curculionidae, Dermestidae, Laemophloeidae, Silvanidae and Tenebrionidae.

The structure of beetles

To identify beetles of stored products a basic understanding of the structure of beetles is needed. Terms illustrated below (Figures C1 and C2) are widely used in the keys and descriptions in this chapter.

C1

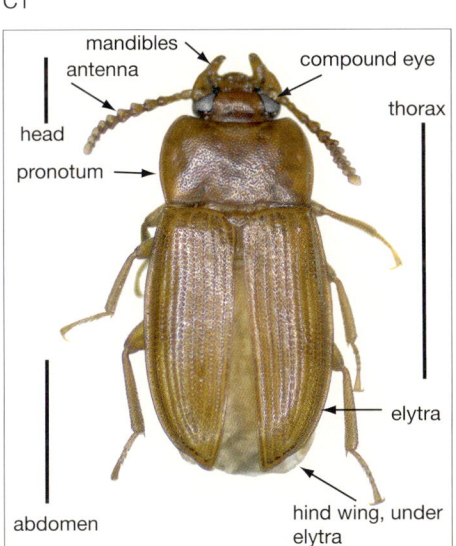

C2

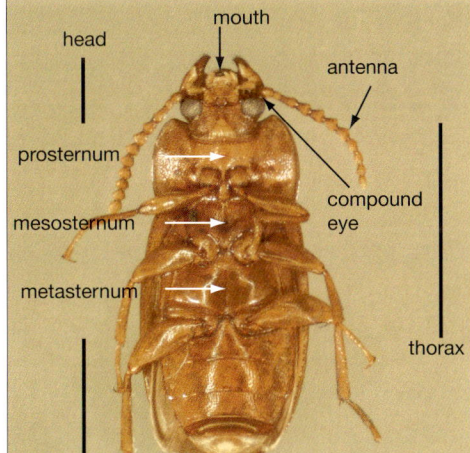

Figure C. Major body components of an adult beetle

Like all insects, the bodies of beetles are divided into three main parts: the head, thorax and abdomen.

Head

On the head are the antennae and eyes and the mouth, which in beetles is designed for biting and chewing. The eyes on each side of the head are compound eyes, so named because they are made up of many ocelli or eye units. In some species a single eye (median ocellus) is also present on the middle of the forehead (Figure 69) (note that it not present in the species illustrated above, *Sitophagus hololeptoides*).

Thorax

The thorax consists of three sections: (from front to back) the prothorax, mesothorax and metathorax. Front, mid and hind legs emerge from each section respectively. Modified hardened fore wings, called the elytra, are attached to the mesothorax. These protect and cover the delicate membranous hind wings that are attached to the metathorax and used for flight.

From above, usually all of the thorax that is visible is the top of the prothorax, known as the pronotum (Figure C1). Most of the top of the mesothorax (the mesonotum) and all of the top of the metathorax (the metanotum) are concealed by the elytra. From below, all three sections of the thorax are visible and are known as the pro-, meso- and metasternum (Figure C2).

Abdomen

The abdomen contains the organs for reproduction and most of the gut. In the side of each segment of the abdomen and thorax is a hole or spiracle through which the insect breathes. In most beetles the upper surface of the abdomen is completely covered by the elytra.

Legs

Legs of beetles are frequently ornamented with structures that are of use in identification. They follow more or less the same plan as legs of all insects; a hind leg of a beetle (*Sitophagus hololeptoides*) is shown as an example (Figure D). The tarsus functions as a foot and the terminal segment of the tarsus often has claws for grip. The tibia and femur operate as the upper and lower leg respectively. The coxa and trochanter form the junction of the leg with the thorax.

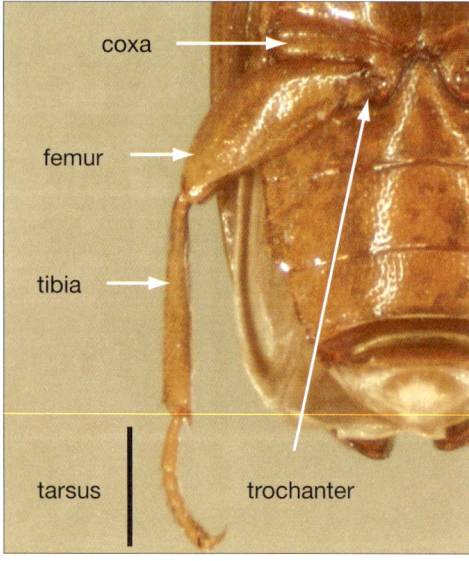

Figure D. Structure of a beetle leg

Identifiying families of beetles of stored products

Adult beetles

Families of beetles can be identified using Figure E or the Key to adult beetles (on page 14).

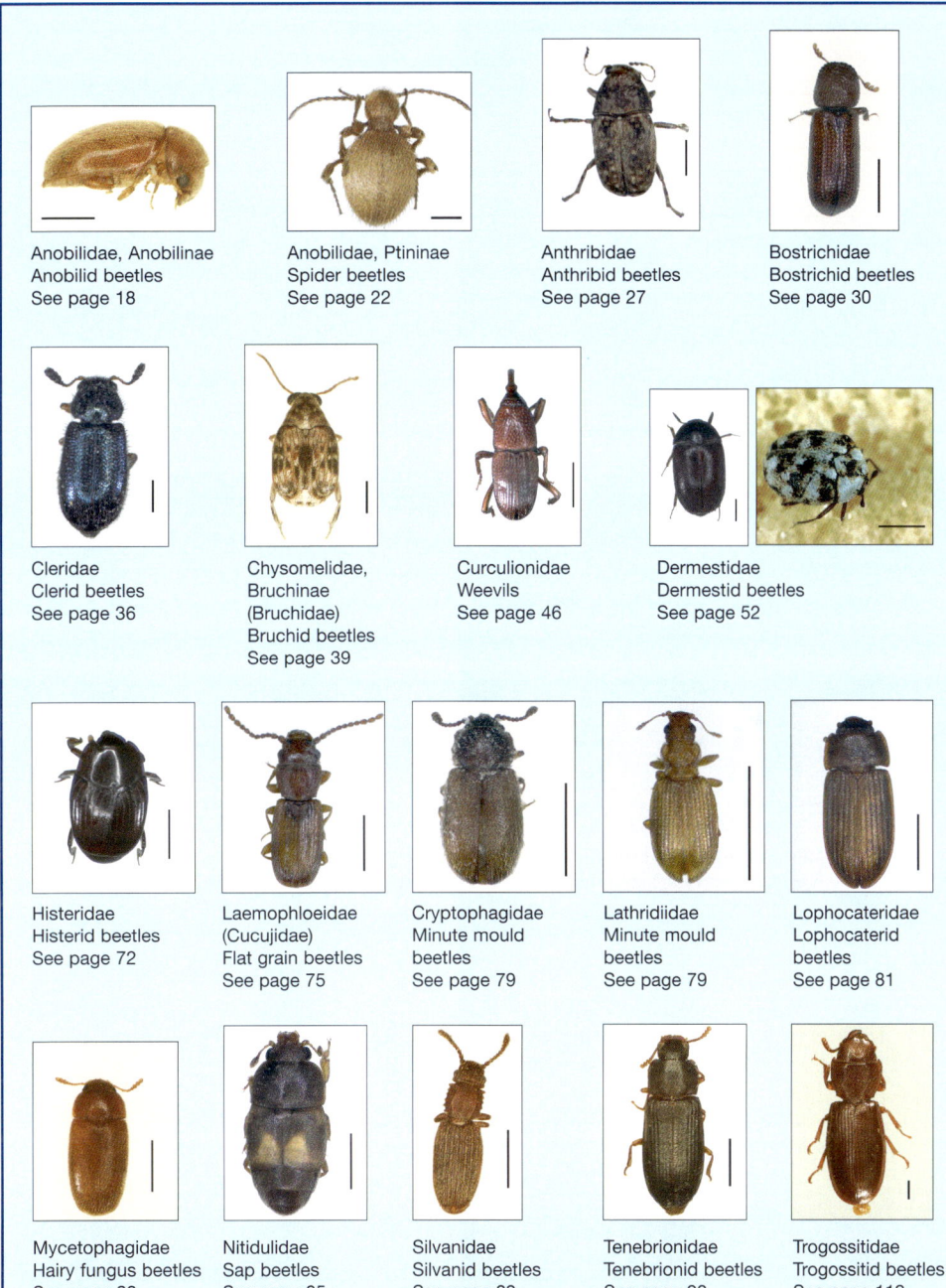

Anobilidae, Anobilinae
Anobilid beetles
See page 18

Anobilidae, Ptininae
Spider beetles
See page 22

Anthribidae
Anthribid beetles
See page 27

Bostrichidae
Bostrichid beetles
See page 30

Cleridae
Clerid beetles
See page 36

Chysomelidae,
Bruchinae
(Bruchidae)
Bruchid beetles
See page 39

Curculionidae
Weevils
See page 46

Dermestidae
Dermestid beetles
See page 52

Histeridae
Histerid beetles
See page 72

Laemophloeidae
(Cucujidae)
Flat grain beetles
See page 75

Cryptophagidae
Minute mould
beetles
See page 79

Lathridiidae
Minute mould
beetles
See page 79

Lophocateridae
Lophocaterid
beetles
See page 81

Mycetophagidae
Hairy fungus beetles
See page 83

Nitidulidae
Sap beetles
See page 85

Silvanidae
Silvanid beetles
See page 89

Tenebrionidae
Tenebrionid beetles
See page 93

Trogossitidae
Trogossitid beetles
See page 119

Figure E. Families of beetles associated with stored products

The following simplified key is designed to separate adult specimens of beetle families covered in this guide. It covers major pest families and others frequently found, but is not a comprehensive key to all possibilities. For a more comprehensive key to families and major genera of stored product Coleoptera, see Halstead (1986) or Kingsolver (1987).

Key to adult beetles associated with stored products

1   Head of beetle with obvious snout, and elbowed antennae (Figures 61, 63) . . . .
. . . . **Weevils – Curculionidae – Pages 46–51 (Figures 56–67)**
Head without obvious snout **2**

2   Spider-like in appearance with very long legs (Figures 7, 8, 10, 12, 141) **3**
Shaped otherwise **4**

3   Body 2–4.5 mm long, colour variable, species range from hairless to very hairy . . . . . .
. . . . **Spider beetles – Anobiidae, Ptininae – Pages 22–26 (Figures 7–20)**
Body 25–30 mm long, hairless and glossy black in colour . . . .
. . . . *Blaps* **spp. – Tenebrionidae (in part) – Pages 97–98 (Figure 141)**

4   From above, elytra are short and leave one or more abdominal segments uncovered
(Figures 21, 39, 105 124) **5**
From above, elytra are long and cover abdomen completely* **8**

5   Body covered in short hairs, elytra richly patterned, antennae long, without spherical club at tip, legs long (Figures 21, 41) **6**
Body hairless, antennae with spherical club at tip, legs relatively short (Figures 105,126) .**7**

6   Head relatively narrow relative to pronotum, antennae long with segments simple or saw tooth like, segments at tip do not form club. Associated with pulses, peas and beans . . .
. . . . **Bruchid beetles – Chysomelidae, Bruchinae – Pages 39–46 (Figures 36–55)**
Head relatively wide relative to pronotum, antennae long with last three segments of antennae somewhat thickened to form loose club. Not associated with pulses, peas and beans . . . . **Anthribid beetles – Anthribidae – Pages 27–30 (Figure 21)**

7   Body very shiny black or metallic in appearance, final segment of antennae much enlarged to form spherical club . . . .
. . . . **Histerid beetles – Histeridae – Pages 72–75 (Figures 105–107)**
Body brown to black, not metallic, in many species elytra often marked with one or two yellow, orange or red spots, antennal club spherical but formed from last three segments
. . . . **Nitidulid beetles – Nitidulidae – Pages 85–88 (Figures 124–127)**

8   When viewed from above, head hidden from view, and held facing downwards, beetles either globular or cylindrical in form (Figures 1, 2, 25) **9**
When viewed from above, head held facing forwards and not hidden from view, body of beetle flattened in form, most species are parallel sided or slightly oval (Figures 68, 108, 133, 171) **10**

9   Body cylindrical in form, hairless, pronotum with rows of blunt forward facing teeth-like structures. Antennae short with last three segments enlarged to form club . . . .
. . . . **Bostrichid beetles – Bostrichidae – Pages 30–36 (Figures 22–32)**
Body globular in form, covered in very short fine hairs, pronotum smooth, antennae long . . . . **Anobiid beetles – Anobiidae, Anobiinae – Pages 18–21 (Figures 1–6)**

10 Sides of beetle adorned with stiff outward facing bristles, species usually seen in storage environments about 5 mm long and in whole or part metallic blue or blue-green in colour . . . .

. . . . Clerid beetles – Cleridae – Pages 36–38 (Figures 33–35)
Body form not as above 11

11 Sides of pronotum adorned with teeth-like structures (Figures 114, 128–134) 12
Side margins of pronotum smooth and not adorned with teeth-like structures 13

12 Sides of pronotum adorned with six teeth-like structures (Figure 132) or front or all corners each adorned with a tooth-like structure (Figure 127). Beetles highly active, flattened and parallel sided, not hairy . . . .
. . . . Silvanid beetles – Silvanidae – Pages 89–93 (Figures 128–135)
Pronotum thickened at front corner and adorned with a tooth-like structure half-way down. Beetles flattened, oval, some hairy . . . .
. . . . Minute mould beetles – Cryptophagidae – Pages 79–81 (Figure 114)

13 Pronotum with raised line parallel with each side (Figure 109), head and pronotum large (Figure 108) taking up about half length of body, body parallel-sided and highly flattened. Antennae hair-like long and unclubbed, up to length of body . . . .
. . . . Cryptolestes spp. – Laemophloeidae – Pages 75–78 (Figures 108–113)
Not as above 14

14 Tiny beetles – body length usually less than 2 mm, tarsi of legs each with no more than 3 segments. Body form variable, some species hairy, elytra of other heavily pitted. Base of abdomen sometimes wider than preceding thorax. Some species appear to have a 'waist' or constriction in pronotum (Figures 115, 118), antennae with 1–3-segmented club . . .
. Minute mould beetles, Plaster beetles – Latridiidae – Pages 79–81 (Figures 115–118)
Larger beetles, body length 2–12 mm long, tarsi of legs each with more than 3 segments. Body shape oval or parallel-sided, antennal club where present has at least of 3 . . . 15

15 Beetle hairy or covered in scales, some species patterned and multicoloured (Figures 71, 68, 94,122) 16
Beetle hairless, not marked with coloured patterns (Figures 119, 171, 185). 18

16 Head with median ocellus present (Figure 69) . . . .
. . . Dermestid beetles – Dermestidae (in part) – Pages 52–72 (Figures 68–79, 93–104)
Median ocellus absent 17

17 Body length 5–10 mm, oval in shape, upperside of beetle black or black with base of elytra brown, underside of abdomen covered in silver hairs with black patches at margins or covered with dark brown hair. Usually associated with material of animal origin . . . . . . Dermestes spp. – Dermestidae (in part) – Pages 52–72 (Figures 80–92)
Body length 2–3 mm, oval in shape, upperside light brown, elytra with parallel lines of hairs, elytra of some species with lightly coloured band. Usually associated with damp material of plant origin . . . .
. . . . Hairy fungus beetles – Mycetophagidae – Pages 83–85 (Figures 121–123)

18 Beetle glossy black, prothorax and elytra separated by 'waist', body length 5–11 mm, flattened, parallel-sided . . . .
. . . Cadelle (Tenebroides mauritanicus) – Trogossitidae – Pages 119–120 (Figure 185)
Prothorax and elytra not separated by waist,* colour pale brown to black, body length 2.5–12 mm, shape flattened oval to parallel-sided 19

19 Length 2.5–12 mm. Front and mid legs each with 5 segmented tarsi, hind legs with 4 segmented tarsi. Flattened parallel-sided beetles (Figure 171) (except Alphitobius spp., Figure 136, which are flattened and oval in shape). From side, eye divided by extension of head (Figure 172) (except Palorus spp., Figure 158), sides of pronotum not expanded as 'flange' . . . .
. . . . Tenebrionid beetles – Tenebrionidae (in part) – Pages 93–119 (Figure 136–184)

Length 2.7–3.2 mm, legs each with 5 segmented tarsi (basal segments difficult to see, so it looks like legs have 4 segmented tarsi), sides of pronotum and elytra flattened as flange (Figures 120). Highly flattened, elytra with distinct longitudinal ridges (Figure 119) . . . .
. . **Siamese grain beetle** (*Lophocateres pusillus*) – **Lophocateridae** – **Pages 81–83 (Figures 119–120)**

Note: * Specimens stored in alcohol can with time become distended. This can lead to the abdomen becoming swollen and extending beyond elytra and the thorax and abdomen becoming distended to form a 'waist'. Tissue distended in this way is soft and white in colour, unlike the hardened pigmented external surfaces of the specimen.

Larvae of beetles

Identification of beetle larvae is relatively difficult and not often required as larvae usually occur in the presence of more easily identifiable adult beetles. Larvae of families of beetles found in stored products have one of five distinct body forms (Figure F).

The body form of larvae of families of beetles found in stored products are given in the table below.

Larval forms of families of beetles frequently associated with stored products

Family	Apodous	Scarabform	Campodeiform	Elateriform	Eruciform	Notes	Figures
Anobiidae			X			legs fully developed	5, 9
Anthribidae			X			legs poorly developed	
Bostrichidae		X				legs fully developed	30
Chysomelidae		X				legs poorly developed	51
Cleridae			X			body mottled pink on light background, hairy	
Cryptophagldae			X			body pale/translucent, flattened	
Curculionidae	X					no legs	65
Dermestidae					X	larvae oval or elongate, species very 'hairy' some with characteristic hastisetae (Fig. 104)	70, 76, 84, 103
Histeridae			X			head with large forward facing mandibles	106
Laemophloeidae			X			body pale/translucent, highly flattened	110
Latridiidae			X			body pale/translucent, flattened	
Lophocateridae			X			body pale/translucent, flattened	
Mycetophagidae			X			body pale/translucent, flattened	
Nitidulidae			X			body pale/translucent, flattened	
Silvanidae			X			body pale/translucent, highly flattened	135
Tenebrionidae				X		body yellowish to dark brown – integument leathery	150, 176
Trogossitidae			X			thorax with obvious dorsal dark coloured areas	

Free-living larvae of moths (Lepidoptera) can be confused with campodeiform beetle larvae. Moth larvae are more cylindrical in form and, unlike beetle larvae, have a pair of false legs (pro-legs) on four segments of the abdomen (see page 122). Unlike beetles, infestations of moth larvae typically produce lots of silken webbing.

Unlike other moths larvae associated with stored products, abdominal pro-legs are virtually absent on larvae of the moth *Sitotroga cerealella* (page 126), so these larvae are thus confusable with those of some beetles. The larvae of *S. cerealella* live internally within grains in a similar manner to larvae of weevils and bostrichid beetles. Unlike the larvae of *S. cerealella,*the larvae of weevils do not possess thoracic legs, while the thorax of mature bostrichid larvae is more enlarged by comparison to larvae of *S. cerealella*.

A comprehensive key to larvae of beetles associated with stored products is provided in Anderson (1987).

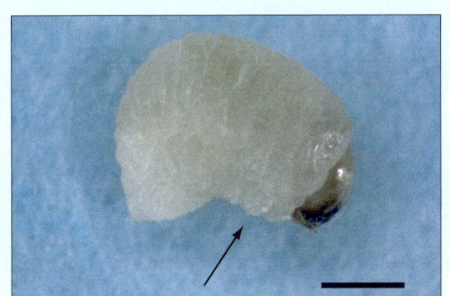

Apodous: legless, immobile, lives internally within foodstuff.

Scarabaeiform: effectively immobile when mature, develops into characteristic 'C' shape as it matures, thorax enlarged relative to abdomen, legs partly or full developed. Lives internally within foodstuff for all or all but initial stage of life.

Left: **Campodeiform:** elongate flattened body with well developed legs, cuticle usually very pale and translucent, head capsule and/ or last abdominal segment may be pigmented. Very active, lives freely within commodity.

Elateriform: body is long and cylindrical, cuticle leathery, legs relatively short. Active, lives freely within commodity.

Eruciform: larvae oval or elongate, species very 'hairy' some with characteristic hastisetae. Active, lives freely within commodity.

Figure F. Larval forms of beetles

Anobiid beetles, Spider beetles
(Family: Anobiidae)

Introduction to family Anobiidae

Two subfamilies of the Anobiidae are associated with stored products: the Anobiinae and the Ptininae. Most Anobiinae are wood borers and many species are serious pests of timber. The Ptininae are scavengers and are associated with dried material of animal and vegetable origin and are frequent inhabitants of animal nests. Members of this subfamily were formerly classified as the family Ptinidae.

Adults of the two subfamilies are physically quite different from each other. Members of the Anobiinae are globular or cylindrical beetles. The head is typically held downwards and from above is largely or completely concealed by the pronotum. The Ptininae are commonly known as spider beetles on account of their general appearance, having globular bodies and long legs and antennae.

Anobiid beetles (Family: Anobiidae, Subfamily: Anobiinae)

Lasioderma serricorne	Cigarette beetle
Stegobium paniceum	Drugstore beetle

Summary

Feeding strategies	primary pest, secondary pest
Commodities attacked	dried material of animal and vegetable origin e.g. tobacco, nuts, herbs and spices, grain and grain products
Distribution	worldwide
Economic importance	high – in processing /retailing industries
Eggs	laid in crevices and folds of commodity
Larvae	scarabeiform, immobile when mature, internal feeders
Adults	short lived, do not feed on commodity, fly readily

Introduction

Most Anobiinae are woodborers and many species are serious pests of timber. Two species, *Lasioderma serricorne* and *Stegobium paniceum*, are frequently found infesting a wide range of stored products.

Identification

L. serricorne (Figures 1–2) and *S. paniceum* (Figures 3–4) are oval, globular beetles 3–4 mm long. Both possess long antennae which are waved rapidly when walking. When disturbed, individuals often 'play dead' by remaining motionless and curling up. Larvae are scarabaeiform with fully functional legs (Figure 5).

Figure 1 *Lasioderma serricorne*, adult, live

Figure 2 *Lasioderma serricorne*, adult, side view, showing antennae and body covered in fine hairs

Figure 4 *Stegobium paniceum*, head from underside, showing antennae

Figure 3 *Stegobium paniceum*, adult, live, showing antennae and longitudinal ridges marked with fine hairs on elytra

Figure 5 *Lasioderma serricorne*, larva, with fully functional legs

Identification of Anobiinae associated with stored products

Antennae long, segments serrate (saw like), elytra, smooth with fine hairs (Figure 2) . . . . *Lasioderma serricorne*
Antennae long, last three segments form loose club, elytra with fine longitudinal striae (ridges) and fine hairs (Figure 4) . . . . *Stegobium paniceum*

Life cycle

Eggs are laid singly, in crevices or folds of the substrate. About 100 eggs are typically laid over a lifespan of about 25 days. Newly hatched larvae cannot attack undamaged grain but they will investigate cracks and crevices in commodities in search for an entry point. The larvae burrow into the foodstuff, becoming more crescent-shaped and immobile as they mature. When ready to pupate they make a flimsy cocoon in the foodstuff. On emergence, the adult spends a few days within the cocoon before biting its way out. Adults are short lived, feed little or not at all. They run quickly and fly readily and well.

Physical limits and optimum rate of multiplication

A low minimum temperature at which development is possible allows *S. paniceum* to breed under cool temperate conditions. The mortality and duration of larval development varies considerably with commodity. Development is slow and mortality is high on material with a low nutritional content, e.g. spices when compared with cereal-based products.

Species	Conditions within which breeding takes place	Shortest development period, with optimum conditions	Maximum monthly rate of increase
Lasioderma serricorne	20–38°C, r.h. > 25%	26 days at 30°C, 70% r.h.	20
Stegobium paniceum	15–34°C, r.h. > 35%	40 days at 30°C, 60–90% r.h.	8

Economic importance

Both species are important pests of a wide range of animal and/or plant origin including dried herbs and spices and high-value processed and pre-mixed products. In addition, *L. serricorne* is an important pest of cured tobacco. Both species are often encountered as pests in manufacturing, retail and domestic situations. In grain storage they are relatively unimportant as pests and are usually associated with residues.

Type of damage and symptoms

Larvae burrow into the foodstuff leaving irregular holes. Infested material becomes contaminated with pupal cocoons and the dead bodies of short-lived adults (Figure 6). Young larvae are able to locate and enter very small holes in packaging. Mature larvae and adults upon emergence will readily chew through packaging material such as plastic, paper, foil laminate and wooden containers leaving neat round holes. Such holes may be very numerous following a heavy infestation.

Ecology

In nature, *S. paniceum* has been found living in bee hives, feeding on pollen collected by bees.

Figure 6 *Lasioderma serricorne*, infestation of dry cat food

Monitoring

Adult *L. serricorne* and *S. paniceum* are attracted to light and are often monitored using light traps. Pheromone-based trapping systems are commercially available for both pests. Acoustic detectors are commercially available which detect the sounds that internal stages of these beetles make during feeding. X-ray photography has been used to detect hidden larvae developing within foods.

Geographical distribution

Both species are found worldwide. *L. serricorne* is not cold hardy and survives in temperate areas only in heated buildings. *S. paniceum* is more frequently encountered in temperate areas, and is cold hardy and can survive unaided winter conditions in temperate regions.

Species	Pest status	USA & Canada	Central & South America	Europe & N.Asia	Mediterranean basin	Africa	S. & SE. Asia	Australia & Oceania
Lasioderma serricorne	●●●	X	X	X	X	X	X	X
Stegobium paniceum	●●●	X	X	X	X	X	X	X

Pest status: ● minor to ●●●● major pest
X: recorded

References

Aitken (1975), Arbogast (1991), Haines (1991).

Spider beetles (Family: Anobiidae, Subfamily: Ptininae – formerly Family: Ptinidae)

Gibbium aequinoctiale	Smooth spider beetle
Gibbium psylloides	Smooth spider beetle, Hump beetle
Mezium affine	Shiny spider beetle, Northern spider beetle
Mezium americanum	Black spider beetle
Niptus hololeucus	Golden spider beetle
Ptinus clavipes	Brown spider beetle
Ptinus fur	Whitemarked spider beetle
Ptinus ocellus (*P. tectus*)	Australian spider beetle
Ptinus raptor	Eastern spider beetle
Ptinus villager	Hairy spider beetle
Pseudeurostus hilleri	
Tipnus unicolor	
Trigonogenius globulus	Globular spider beetle

Summary

Feeding strategies	secondary pest, scavenger
Commodities attacked	dried material of animal and vegetable origin
Distribution	worldwide especially temperate regions
Economic importance	low–medium
Eggs	laid in crevices and folds of commodity
Larvae	scarabaeiform, immobile when mature
Adults	long lived, feed, some species can fly

Introduction

Spider beetles are so named because of their superficial resemblance to small spiders. In nature, they are scavengers and can be found under bark, in rotting logs, caves and in animal nests. A number of genera are associated with stored produce as minor pests and scavengers. They are most often encountered in temperate regions.

Identification

Spider beetles are long-legged globular beetles 2–4.5 mm long with a superficial resemblance to a small spider (Figures 7–20). Antennae are long and hair-like. Larvae are scarabaeiform with fully functional legs (Figure 9). Adults of major genera can be separated by characteristics below.

Figure 7 *Gibbium psylloides*, adult, showing antennae and distinctive glossy red appearance

Figure 8 *Mezium americanum*, adult, showing antennae and distinctive glossy black elytra

Figure 9 *Mezium americanum*, larva, with fully functional legs

Figure 10 *Niptus hololeucus*, adult, showing long antennae and body covered in light brown hairs

Figure 11 *Niptus hololeucus*, adult, head, showing wide raised area between base of antennae

Figure 12 *Ptinus clavipes*, adult

Figure 13 *Ptinus clavipes*, adult, showing long antennae, elytra uniform brown with dark brown hairs

Figure 14 *Ptinus clavipes*, adult, head, showing narrow raised area between base of antennae

Figure 15 *Ptinus clavipes*, adult, hind leg, showing relatively short trochanter

Figure 16 *Ptinus fur*, adult, showing white patches on elytra

Figure 17 *Pseudeurostus hilleri*, adult, showing thorax and elytra with short golden hairs

Figure 18 *Pseudeurostus hilleri*, adult, hind leg, showing relatively long trochanter

Figure 19 *Tipnus unicolor*, adult, thorax and elytra with brown or golden hairs

Figure 20 *Trigonogenius globulus*, adult, showing elytra with dark patches and covered with hairs of variable length

Key to major genera of Ptininae associated with stored products

1   Elytra shiny and hairless (Figures 7–8) 2
    Elytra not shiny and covered in hairs (Figures 10, 12–13) 3

2   Pronotum and elytra shiny and hairless, colour brownish red to black (Figure 7) . . . . .
    . . . . *Gibbium* spp.
    Pronotum covered with golden hair, elytra shiny black and hairless (Figure 8) . . . .
    . . . . *Mezium* spp.

3   Elytra with patches of dark brown and black hairs amongst golden hairs, above which
    are erect golden hairs of variable length arranged in longitudinal rows (Figure 20) . . .
    . . . . *Trigonogenius globulus*
    Elytra not so marked 4

4   Width of raised area between base of antennae is wide, at least half length of 1st antennal
    segment (Figure 11) 5
    Width of raised area between base of antennae is narrow, less than quarter of length of
    1st antennal segment (Figure 14) 6

5   Elytra with brown or golden hairs, elytra surface not obscured from view by hairs
    (Figure 19) . . . . *Tipnus unicolor*
    Pronotum and elytra covered with mixture of recumbent (laying) and erect yellow hairs,
    (Figure 10) . . . . *Niptus hololeucus*

6   Pronotum and elytra with brown or golden hairs. Trochanter of hind leg short. (Figure
    15). Some distinctive species: *Ptinus fur* – elytra have white patches *Ptinus ocellus* – fresh
    specimens densely covered with brownish hairs so that elytra surface not visible (Figures
    12, 13, 16) . . . . *Ptinus* spp.
    Pronotum and elytra with golden hairs (Figure 17). Trochanter of hind leg long
    (Figure 18) . . . . *Pseudeurostus hilleri*

As well as the species listed above, there are others which are occasionally found in human habitation. For further details, refer to the listed references.

Life cycle

Eggs are laid on the surface of bulk commodities or inserted in crevices or between the weave of sacks commodities. In bulk commodities, infestations remain near the surface. Larvae spin mucous feeding shelters. When ready to pupate they spin tougher cocoons. Adult spider beetles are usually most active at night.

Two forms of *Ptinus clavipes* exist – a diploid sexual and a triploid parthenogenetic form. These differ in appearance and as a result were once thought to consist of two separate species. Males of *Ptinus* spp. are smaller and less globular in appearance than females. Males of most *Ptinus* species can fly, females cannot.

Physical limits and optimum rate of multiplication

Compared to many other storage pests, ptinids are very tolerant of low temperatures and as a result are most often seen in temperate regions. Spider beetles breed slowly but are long-lived. Species of temperate origin such as *Ptinus* species have resting stages that restrict population development to one generation a year. Species originating in warmer regions, for example *Mezium* and *Gibbium* spp., are capable of undertaking two or three generations a year.

Species	Conditions within which breeding takes place	Shortest development period, with optimum conditions	Maximum monthly rate rate of increase
Gibbium aequinoctiale	20–35°C, 40–70% r.h.	45 days at 33°C	
Gibbium psylloides	> 20°C, > 30% r.h.	31–34°C	4
Mezium affine	20–33°C, 30–70% r.h.	62 days at 29–33°C	2.5
Niptus hololeucus	> 10–25°C, >50% r.h.	96 days at 20°C	2
Ptinus clavipes	> 19°C, > 50% r.h.	21–27°C	1+
Ptinus fur	> 10°C	21–25°C	2
Ptinus ocellus (P. tectus)	10–28°C, > 70% r.h.	61 days at 27°C	4
Pseudodeurostus hilleri	< 28°C, 40–70% r.h.	53 days at 20°C	
Tipnus unicolor	> 12–25°C, > 60% r.h.	120 days at 18°C	
Trigonogenius globulum	< 28°C	24°C	

Economic importance

Spider beetles are minor pests typically associated with residues in a storage structure. They mostly breed in damp corners, crevices, animal nests etc. where residues, faeces and dead animals may build up. Presence of large numbers of spider beetles is usually a sign of poor sanitation. Full grown larvae may damage solid materials such as wooden storage structures, by burrowing into them when preparing to pupate.

An ability of these insects to breed at low temperatures has allowed them to become relatively important storage pests in cool temperate regions such as Canada and northern Europe.

Type of damage and symptoms

When present in large quantities, larvae of spider beetles may mat the surface of the infested commodity together with mucous feeding shelters and pupal cocoons.

Ecology

In nature, spider beetles are found in birds' nests, caves, rotten wood and deposits of damp material of animal and vegetable origin.

Monitoring

Ptinids are often nocturnal in their activity. During daylight hours they often hide in crevices. Crevice traps made from sacking material or cardboard and searching at night can be utilised to locate these insects.

Geographical distribution

Species	Pest status	USA & Canada	Central & South America	Europe & N.Asia	Mediterranean basin	Africa	S. & SE. Asia	Australia & Oceania
Gibbium aequinoctiale	●	X	X	X	X	X	X	X
Gibbium psylloides	●			X	X			
Mezium affine	●	X		X	X			
Mezium americanum	●	X	X	X	X	X	X	X
Niptus hololeucus	●	X	X	X	X		X*	X
Ptinus clavipes	●	X	X	X	X	X		X
Ptinus fur	●●	X	X	X	X		X	X
Ptinus tectus (P. ocellus)	●●	X		X	X		X*	X
Ptinus raptor	●	X		X				
Ptinus villager	●	X		X				
Tipnus unicolor	●	X		X				
Sphaericus gibboides	●	X		X	X			
Trigonogenius globulus	●	X	X	X		X*		X

Pest status: ● minor to ●●●● major pest
X: recorded
*: temperate regions only

Gibbium aequinoctiale and *Mezium americanum* are most frequently encountered in warm climates. The many species of *Ptinus* and the other genera listed are mostly encountered in temperate regions.

References

Aitken (1975), Bousquet (1990), Haines (1991), Hinton (1940), Howe (1991), Mound (1989), Spilman (1987a).

Anthribid beetles
(Family: Anthribidae)

Araecerus fasciculatus	Cocoa weevil, Coffee-bean weevil

Summary

Feeding strategy	primary pest
Commodities attacked	cocoa, coffee, dried cassava and yams, maize, groundnuts, brazil nuts, spices especially nutmegs
Distribution	tropical
Economic importance	high on high value crops
Eggs	laid onto seed or root
Larvae	immobile, concealed within commodity
Adults	do not feed on commodity, can fly

Introduction

The Anthribidae is a family of beetles most of which feed on fungi and dead wood. Only one species *Araecerus fasciculatus* is well known as a pest of stored products.

Identification

Figure 21 *Araecerus fasciculatus*, adult, showing antennae, head capsule wide relative to width of pronotum, patterning on elytra

A. fasciculatus is a globular beetle 3–5 mm long with long legs and long antennae (Figure 21) and is somewhat similar in appearance to bruchids. Elytra are patterned with small light and dark patches to give a chequered appearance. Elytra are short and leave the last segment of the abdomen exposed. The last three segments of the antennae are somewhat thickened and form a loose club. The larvae are scarabaeiform and hairy, with only vestigial legs.

Adult *A. fasciculatus* can be confused with bruchid beetles (see page 40). Compared to bruchids of similar size, the head capsule of *A. fasciculatus* is much wider relative to the width of the pronotum. Antennae are also different. Anthribids and bruchids attack quite different commodities and are unlikely to be found together.

Life cycle

Eggs are laid on the surface of the infested commodity. Larvae bore into the food material where they remain until adulthood. When infesting coffee cherries, larvae first feed on the fruit pulp before attacking the seed. Adult beetles can fly well and infestation of commodity may occur before harvest. Infestation of coffee may begin before harvest or during the initial phase of the drying process and will continue until the commodity is dried to below about 8% moisture content.

Physical limits and optimum rate of multiplication

Species	Conditions within which breeding takes place	Shortest development period, with optimum conditions	Maximum monthly rate of increase
Araecerus fasciculatus	> 22°C, r.h. > 60%	26–66 days at 28–32°C, > 60% r.h.	40

A. fasciculatus breeds most rapidly and successfully under conditions of high humidity. Drying of the commodity greatly increases mortality and the length of time taken to develop to adulthood.

Economic importance

A. fasciculatus can attack a wide range of commodities but is best known as a pest of coffee, cocoa and spices such as nutmeg. It will also attack some tropical nuts and dried roots such as yams and cassava. It is occasionally found attacking maize.

A. fasciculatus is an important pest of high value beverage crops and spices. On dry commodities kept in good condition damage is minimal, but if crops are stored damp in poor conditions then damage caused can become significant. Infestation of other materials listed is most often encountered under conditions of tropical subsistence agriculture.

Type of damage and symptoms

Emergence of adult beetles results in the excavation of circular emergence holes in the commodity and large cavities within seeds.

Ecology

Initial infestation by this insect is known to provide access for other pests, notably moths of the genera *Cadra* and *Ephestia*.

Monitoring

Infestations of these insects quickly become obvious as a result of the presence of the active adults and emergence holes in infested commodities. Acoustic detectors are commercially available which detect the sounds that internal stages of these beetles make during feeding. X-ray photography has been used to detect hidden larvae developing within grains.

Geographical distribution

Species	Pest status	USA & Canada	Central & South America	Europe & N.Asia	Mediterranean basin	Africa	S. & SE. Asia	Australia & Oceania
Araecerus fasciculatus	●●		X			X	X	X

Pest status: ● minor to ●●●● major pest
X: recorded

A. fasciculatus is established in tropical regions where its crop hosts are grown. For example in Australia, it is restricted to the tropical north-east. In temperate regions, it may be intercepted on imported beverage crops and spices.

References

Aitken (1975), Arbogast (1991), Haines (1991).

Bostrichid beetles
(Family: Bostrichidae)

Dinoderus bifoveolatus	West African ghoon beetle
Dinoderus brevis	
Dinoderus distinctus	
Dinoderus japonicus	Japanese ghoon beetle
Dinoderus minutus	Bamboo false powder post beetle
Dinoderus ocellaris	
Prostephanus truncatus	Larger grain borer
Rhyzopertha dominica	Lesser grain borer

Summary

Feeding strategy	primary pest
Commodities attacked	whole cereal grains, dried root crops, bamboo, rattan, wood
Distribution	worldwide, but *P. truncatus* restricted to Africa and southern USA to northern South America
Economic importance	*R. dominica, P. truncatus* – high, *Dinoderus* spp. – low–medium
Eggs	laid on grain or in tunnels bored by adults
Larvae	scarabaeiform, immobile when mature, live within commodity
Adults	long lived, feed on commodity, can fly

Introduction

Bostrichid beetles are mainly stem, wood and root borers and as a result many are important timber and forestry pests. However, a few genera have become adapted to feeding on cereal grains and dried root crops. Of these, *Rhyzopertha dominica* and *Prostephanus truncatus* rank among the most important pests of stored cereal grains.

Identification

The bostrichids that attack stored products are dark brown to black beetles 3–5 mm long and are cylindrical in cross-section (Figures 22–32). When viewed from above, the head of the beetle is held bent downwards and is concealed by the pronotum. The pronotum is adorned with many small lumps or tubercles. Larvae are scarabaeiform and have functional legs (Figure 30).

In comparison with other storage beetles, bostrichids are distinctive in appearance. Wood borers of the sub-family Lyctinae (Family Bostrichidae) are of somewhat similar appearance and are occasionally found in stores, usually as structural pests. The antennal club of Scolytinae (Family Curculionidae) is usually in the form of a sphere rather than as a loose three-segmented club, while lyctids have a distinctive two-segmented club (Figure 31).

Other bostrichids are occasionally found in stored commodities, especially in the tropics. Of these, the most commonly encountered are species of *Heterobostrychus* and *Sinoxylon*. These feed primarily on timber but may occasionally attack dried roots. Many wood-boring bostrichids are larger than the storage species and their prothorax and elytra are often ornamented with very obvious hooks and spines.

P. truncatus can be easily confused with *Dinoderus* spp., especially as both genera can be found together. *R. dominica* is relatively distinct and is not usually found in association with either *P. truncatus* or *Dinoderus* spp. Genera can be separated as below or using Anon (1993) or Haines (1991). Most economic species of the genus *Dinoderus* can be identified using the key of Spilman (1982).

Morphological characters to separate genera of storage bostrichids

Dinoderus spp. (Figures 22–24)
- Length about 3 mm, but broader than *R. dominica*. Colour – dark brown (Figure 22–23).
- Tip of abdomen with rounded corners when viewed from above or below (Figure 22).
- When viewed from the side end of the elytra is rounded, like a quarter of a circle (Figure 23).
- Two oval shaped depressions in pronotum – present in most species but not *D. distinctus* (Figure 24).

Figure 22 *Dinoderus* spp., adult, showing antennae, thorax and rounded tip of elytra

Figure 23 *Dinoderus* spp., adult, side view, showing rounded tip of elytra

Figure 24 *Dinoderus* spp., adult showing pair of depressions at base of thorax

Prostephanus (Figures 25–27)
- Length about 4 mm. Colour – black (Figure 25).
- Tip of abdomen square when viewed from above or below (Figure 25–26).
- Boundary between end and side of elytra marked with a ridge (Figure 25–26).

Figure 25 *Prostephanus truncatus*, adult, showing antennae and square tip of elytra with ridge marking junction of side and tip of elytra

Figure 26 *Prostephanus truncatus*, adult, underside, showing square tip of elytra

Figure 27 *Prostephanus truncatus*, infestation on maize cobs

Rhyzopertha (Figures 28–32)
- Length 3 mm – narrower in cross-section than *Dinoderus* species. Colour – dark reddish brown (Figure 28, 29).
- Tip of abdomen tapered when viewed from above or below (Figure 28).
- When viewed from the side, the ends of the elytra curve gradually (Figure 29).

Figure 28 *Rhyzopertha dominica*, adult, showing antennae, thorax and tapered tip of elytra

Figure 29 *Rhyzopertha dominica*, adult, side view, showing thorax and gradual slope of tip of elytra

Figure 30 *Rhyzopertha dominica*, larva, with fully functional legs

Figure 31 *Lyctus brunneus* (left), detail of antennae with two-segmented club, and *Rhyzopertha dominica* (right), detail of antennae with loose three-segmented club

Figure 32 *Rhyzopertha dominica*, infestation on wheat

Life cycle

Details for *R. dominica* and other species are similar. Eggs are laid singly or in batches of up to about 20 in cracks or crevices in food media or amongst flour and debris produced by adult burrowing. Newly hatched larvae either bore into grains or feed amongst the matrix of damaged grains and flour produced by the adults. Young larvae of *R. dominica* are free living and can search out cracks and weaknesses in grains. As they mature, larvae become increasingly 'C' shaped (scarabaeiform) and more immobile (Figure 30). Larvae are not cannibalistic, and high densities will co-exist. In the case of *P. truncatus*, 8–10 larvae can successfully develop within a single maize kernel. Pupation occurs within a grain or in the matrix of debris and flour. Adults are long-lived, can fly well and feed voraciously. Some 400 eggs can be laid under optimal conditions by a female over a life span of 3 months or more.

Physical limits and optimum rate of multiplication

Species	Conditions within which breeding takes place	Shortest development period, with optimum conditions	Maximum monthly rate of increase
Dinoderus spp.		c. 180 days at 35°C, 75% r.h	c. 3.5
Prostephanus truncatus	18–36°C, 40–90% r.h	26 days at 30°C, 75% r.h	25
Rhyzopertha dominica	20–38°C, > 30% r.h.	25 days at 34°C, 70% r.h	20

Both *P. truncatus* and *R. dominica* can breed under a wide range of climatic conditions, but they are more tolerant of hot conditions and dry grain than other major pest species, notably *Sitophilus* spp. (Coleoptera : Curculionidae). Relative to *P. truncatus* and *R. dominica*, the rate of population growth of *Dinoderus* spp. is low.

Economic importance

Dinoderus spp. are best known as pests of goods or structures made from bamboo and rattan. Worldwide the most frequently encountered species is *D. minutus*. *Dinoderus* spp. can be damaging pests of maize cobs and dried cassava stored under conditions of tropical subsistence agriculture. However, they are not usually encountered in well managed commercial storage systems.

R. dominica is a major pest of whole cereals, especially of wheat, barley, sorghum and rice. However, it is much less serious as a pest of maize. It may also attack compacted milled products. Along with *Sitophilus* spp., it ranks as the most important pest of stored cereals worldwide. *R. dominica* is an important pest in all types of grain storage whether traditional, bag or bulk, including the most modern mechanised bulk handling systems.

P. truncatus is a major pest of maize and dried cassava under conditions of subsistence agriculture, especially where maize is stored on the cob. Infestation can occur prior to harvest. Under such conditions, damage caused by this pest is severe. Its recent arrival in Africa has caused considerable additional hardship for many communities. While *P. truncatus* is only a minor pest in bulk or bagged maize in commercial storage, such systems have been implicated in transporting this pest into new areas. As a consequence, *P. truncatus* is specifically targeted by quarantine authorities in many countries to prevent its introduction and / or restrict its spread.

Type of damage and symptoms

Damage caused by these insects to stored commodities is distinctive and heavy (Figures 27, 32). Adults burrow extensively leaving tunnels and irregular-shaped holes, and produce large amounts of flour. Feeding and burrowing by larvae further adds to the damage. Weight loss due to flour production outweighs losses due to direct consumption. Severe infestations of these insects can easily lead to the physical destruction of the commodity, its packaging and even the storage structure if left unchecked. Grain heavily infested with *R. dominica* has to some people a very characteristic sweet odour. Unlike populations of grain weevils (*Sitophilus* spp.), infestations of bostrichids do not appear to cause significant heating and moulding of infested grains. *R. dominica* is known to be highly allergenic and persons repeatedly exposed to this insect or commodities infested by it often develop respiratory complaints.

Ecology

The switch by these wood-boring insects to attack stored grain appears to be a relatively recent development in evolutionary terms. The three main genera appear to be at different points in their development as storage pests. *R. dominica* appears to be most closely associated with stored products and the storage environment but populations in natural habitats are known. Large populations of *P. truncatus* still occur in a wide range of forest and woodland types independent of grain production and storage. Such populations make it effectively impossible to eradicate this species once it becomes established in an area. However, when it attacks maize or cassava, population development is rapid. *Dinoderus* spp. remain mostly pests of timber, vines and bamboo. Populations of *Dinoderus* spp. develop relatively slowly and only become significant pests on commodities stored under poor conditions in tropical environments.

Other bostrichids may develop in time into storage pests. This is most likely to occur where infested or vunerable timber, vines or bamboo is used for the construction of storage structures.

Then there is the potential for these insects to be maintained in close contact with stored commodities, from which development of grain feeding populations could occur.

Adult bostrichids are strong fliers. In the case of *P. truncatus,* large numbers can be attracted from the surrounding areas to a store by aggregation pheromones produced by beetles already there. This behaviour leads to a characteristic patchy distribution of infestations within an area, with many stores unaffected while others nearby become heavily infested. In addition, infestations of *P. truncatus* may begin prior to harvest. Within a grain mass, bostrichids tend to aggregate and move little, often making them difficult to detect. They have difficulty in attacking loose grain, such as the surface of a bulk, owing to their inability to cling to grains while attempting to burrow into them. Populations are most likely to develop where grain is well compacted such as deep within a grain bulk or when grain is held firm such as on a maize cob. Populations of *R. dominica* are severely affected by frequent moving or turning of grain – a means of control which was used in Australia before the widespread adoption of fumigants and grain protectants.

P. truncatus and *R. dominica* often infests grain along with grain weevils *Sitophilus* spp. The bostrichids, however, are more tolerant of hot dry grain than *Sitophilus* spp. and so tend to dominate under such conditions.

Classical biological control of *P. truncatus* in Africa has been attempted by the introduction of a predatory beetle *Teretrius nigrescens* (Coleoptera: Histeridae) from Mexico and central America.

Monitoring

Adult bostrichids can be attracted to flight traps baited with synthetic aggregation pheromones. Specific pheromones blends for *R. dominica* and *P. truncatus* are commercially available and are highly effective. Both species respond poorly to pitfall and probe traps inserted into grain bulks.

Geographical distribution

Species	Pest status	USA & Canada	Central & South America	Europe & N.Asia	Mediterranean basin	Africa	S. & SE. Asia	Australia & Oceania
Dinoderus bifoveolatus	●●	X	X		X	X	X	
Dinoderus brevis	●						X	
Dinoderus distinctus	●●					X		
Dinoderus japonicus	●	X					X	X
Dinoderus minutus	●●	X	X	X	X	X	X	X
Dinoderus ocellaris	●						X	X
Prostephanus truncatus	●●●●	X®	X			X		
Rhyzopertha dominica	●●●●	X	X	X	X	X	X	X

Pest status: ● minor to ●●●● major pest
X: recorded
®: restricted distribution

Dinoderus species occur widely in tropical regions. *D. distinctus* appears to be of African origin, other species are Asian. When intercepted in temperate regions, it is usually as a result of the import of infested goods, such as bamboo products, from tropical regions.

R. dominica occurs worldwide in both temperate and tropical regions. It is a serious pest in tropical to warm temperate climates. It thrives in major wheat producing areas such as Australia, Argentina and parts of the USA. It is unlikely to be a major pest in cool temperate climates unless it is infesting warm grain.

In contrast, *P. truncatus* has a restricted but expanding distribution. Until the late 1970s, it was only known from the Americas, from the extreme south of the USA, through Mexico and Central America to northern South America. In the late 1970s it was accidentally introduced into Tanzania, East Africa. It has since spread into neighbouring countries and by 2001 had reached the northern border of South Africa. A separate accidental introduction also occurred in Togo, West Africa, from where it has spread to neighbouring countries including Ghana, Benin, Nigeria, Burkina Faso and Guinea Conakry. It is perhaps now only a matter of time before it becomes established in all maize and cassava growing areas of sub-Saharan Africa.

References

Aitken (1975), Anon (1993), Arbogast (1991), Haines (1991), Hodges (1986, 1994), Spilman (1982).

Clerid beetles, Chequered beetles
(Family: Cleridae)

Necrobia rufipes	Redlegged ham beetle
Necrobia ruficollis	Redshouldered ham beetle
Necrobia violacia	Blacklegged ham beetle
Thaneroclerus buquati	

Summary

Feeding strategies	secondary pest, predator
Commodities attacked	copra, oilseeds, products of animal origin
Distribution	worldwide – warm temperate to tropical
Economic importance	medium
Eggs	laid in amongst commodity
Larvae	campodeiform, mobile, external feeders
Adults	long lived, feed on commodity, fly readily

Introduction

The Cleridae are mostly predaceous beetles and many are found in forest and woodland environments where they prey on wood- and bark-boring insects. Two genera, *Necrobia* and *Thaneroclerus*, are associated with stored products as both pests and/or predators of other insects.

Identification

Clerids are distinctive flattened parallel-sided beetles about 5 mm long (Figures 33–35). The antennae have a very distinct three-segmented club. The sides of the thorax have stiff, bristle-like hairs pointing outwards. *Necrobia* spp. are unmistakable among stored product insects on account of their metallic blue-green colouration.

Figure 33 *Thaneroclerus buquati*, adult, showing general form and uniform brown colour

Figure 34 *Necrobia ruficollis*, adult, showing antennae, stiff hairs on margin of thorax, bicoloured elytra

Figure 35 *Necrobia rufipes,* adult, showing uniform metallic blue thorax and elytra

Key to clerid beetles associated with stored products

1 Colour of adult uniform dull brown . . . . *Thaneroclerus buquati* (Figure 33)
   Adult with iridescent blue/green colouring 2

2 Thorax and base of elytra reddish brown, rest of elytra metallic blue-green . . . .
   . . . . *Necrobia ruficollis* (Figure 34)
   Upper surface of the head, thorax and elytra all metallic blue-green 3

3 Antennae and legs reddish-brown in colour in clear contrast to rest of body . . . .
   . . . . *Necrobia rufipes* (Figure 35)
   Antennae and legs dark blue-black in colour . . . . *Necrobia violacea*

Life cycle

Eggs of *Necrobia* spp. are laid on the commodity surface. Larvae burrow into the material. When ready to pupate they spin a cocoon in the infested commodity or in a crevice or tunnel bored elsewhere. Adult beetles feed on the commodity and are long lived and under warm conditions will fly readily in search of new food.

Physical limits and optimum rate of multiplication

Species	Conditions within which breeding takes place	Shortest development period, with optimum conditions	Maximum monthly rate of increase
Necrobia rufipes	21–42°C	30–34°C, high humidity	25

Necrobia spp. thrive best under hot humid conditions.

Economic importance

Necrobia rufipes is an important pest of copra and sometimes of oilseeds and cocoa. It also attacks dried fish and other animal products, often in association with *Dermestes* spp. *N. ruficollis* appears to be largely restricted to materials of animal origin. *N. violacea* is occasionally found in stored products, where it may feed on the dead remains of other insects.

Type of damage and symptoms

Adult insects feed on the surface of the infested material while the larvae burrow deeply into it. Infested materials become contaminated with cast skins and cocoons. Larvae may burrow indiscriminately into other materials such as the storage structure to pupate.

Ecology

Clerids are predatory beetles which usually prey on wood- or bark-boring beetles. *Thaneroclerus buquati* is sometimes encountered in stored products as a predator of the cigarette beetle *Lasioderma serricorne*. *Necrobia* spp. while also predatory, are also capable of breeding on material of entirely vegetable origin such as copra and cocoa.

Monitoring

Adult *Necrobia* are mobile insects and are likely to be attracted by food odours. Mesh bags filled with dried fish and/or copra may prove attractive to these insects.

Geographical distribution

Species	Pest status	USA & Canada	Central & South America	Europe & N.Asia	Mediterranean basin	Africa	S. & SE. Asia	Australia & Oceania
Necrobia rufipes	●●	X	X	X	X	X	X	X
Necrobia ruficollis	●●	X	X	X	X	X	X	X
Necrobia violacea	●	X	X	X	X	X	X	X
Thaneroclerus buquati	P		X		X	X	X	X

Pest status: ● minor to ●●●● major pest
X: recorded
P: predator

T. buquati is widely distributed mainly in tropical regions. *Necrobia rufipes* and *Necrobia ruficollis* are cosmopolitan but are most abundant in the tropics, especially in Asia and Africa. Both are imported from time to time into temperate regions and may survive winter conditions in heated premises. *Necrobia violacea* is found worldwide.

References

Aitken (1975), Arbogast (1991), Haines (1991).

Bruchid or seed beetles

(Family: Chysomelidae, Subfamily: Bruchinae – formerly Family: Bruchidae)

Acanthoscelides obtectus	Bean weevil
Acanthoscelides zeteki	
Bruchus spp.	Pea weevils
Bruchidius spp.	
Callosobruchus analis	
Callosobruchus chinensis	Southern cowpea weevil
Callosobruchus maculatus	Cowpea weevil
Callosobruchus phaseoli	
Callosobruchus rhodesianus	
Callosobruchus subinnotatus	
Callosobruchus theobromae	
Caryedon serratus	Groundnut bruchid
Zabrotes subfasciatus	Mexican bean weevil

Summary

Feeding strategy	primary pest
Commodities attacked	pulses
Distribution	worldwide, especially tropics
Economic importance	high
Eggs	laid or stuck individually onto seed or pod
Larvae	scarabaeiform, immobile, concealed within seed
Adults	can be long lived, do not feed on commodity, fly readily

Introduction

Bruchid beetles attack ripe and ripening seeds. They are especially associated with the seeds of legumes. Species associated with stored products are almost exclusively pests of dried and ripening seeds of legumes and are by far the most important storage pests of these commodities. Bruchids do not attack cereal grain or cereal-based products.

Identification

Bruchids are distinctive globular beetles with long legs and long antennae (Figures 36–55). Species that attack stored products are 3–7 mm in length. The elytra are patterned with light and dark patches, and are short so they do not fully cover the abdomen. The underside of the abdomen is covered in fine hairs. Adults are very active and will readily fly and run rapidly. However, when disturbed they may feign death and remain motionless for minutes.

The general form and appearance of these insects together with their association with pulses make it unlikely that bruchids will be confused with other beetles associated with stored products.

The most reliable feature to identify storage bruchids to genus, and in some cases to species, is by examination of the arrangement of spines and teeth-like structures on the hind leg. The general appearance of specimens is somewhat unreliable as colour patterns are variable and specimens quickly become worn. Specimens of *Bruchus* and especially *Caryedon* are somewhat larger than other species but the remainder are similar in size. Once identified to genus the structure of the male genitalia can be reliably used to identify specimens to species. Adults of most storage species can be identified by the keys of Haines (1989, 1991) and Kingsolver (1987).

Larvae are scarabaeiform and have greatly reduced legs (Figure 51).

Morphological characters to separate genera of storage bruchids

Acanthoscelides (Figures 36–38)
- Inner ridge of ventral margin of hind femur with 3 or 4 'teeth' (Figure 37)

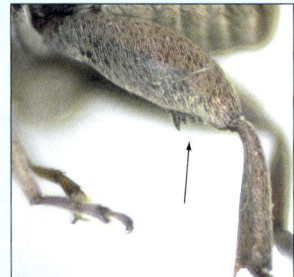

Figure 36 *Acanthocselides obtectus*, adult, live, showing long antennae, patterned elytra and exposed final segments of abdomen

Figure 37 *Acanthoscelides obtectus*, adult, hind leg, spines on ventral margin of hind femur

Figure 38 *Acanthoscelides obtectus*, infestation in kidney beans with emergence holes

Bruchidius (Figures 39–40)
- Inner ridge of ventral margin of hind femur with spine or no spine (Figure 40)

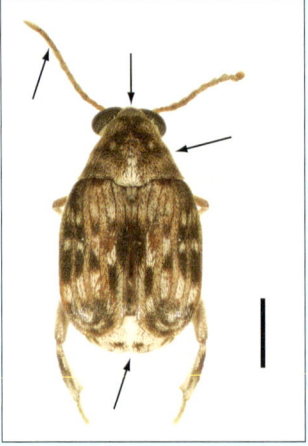
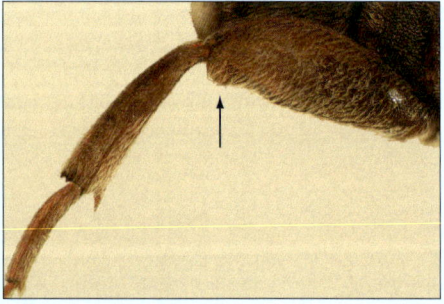

Figure 39 *Bruchidius*, adult, showing antennae, head capsule narrow relative to width of pronotum, short patterned elytra and exposed final segments of abdomen

Figure 40 *Bruchidius* spp., adult, hind leg, spine on ventral side of hind femur

Bruchus (Figures 41–43)

- Inner ridge of ventral margin of hind femur with spine (Figure 42). Side of pronotum with spine (Figure 43)

Figure 41
Bruchus pisorum, adult, live

Figure 42
Bruchus pisorum, adult, spine on ventral side of hind femur of hind leg

Figure 43 *Bruchus pisorum*, adult, showing blunt spine on margin of thorax

Callosobruchus (Figures 44–51)

- Inner and outer ridge of ventral margin of hind femur each with spine (Figure 46, 49)

Figure 44 *Callosobruchus* infestation on dried peas, showing attached eggs and adult emergence holes

Figure 45 *Callosobruchus analis*, adult

Figure 46 *Callosobruchus analis*, adult, hind leg, spines on ventral side of hind femur

Figure 47 *Callosobruchus chinensis*, adult, male with distinctive antennae

Figure 48 *Callosobruchus maculatus*, adult

Figure 49 *Callosobruchus maculatus*, adult, hind leg, spines on ventral side of hind femur

Figure 50 *Callosobruchus phaseoli*, adult

Figure 51 *Callosobruchus*, larva, with greatly reduced legs

Caryedon (Figures 52–53)

- Hind femur enlarged, with one large tooth and 11–12 smaller teeth on ventral margin (Figure 53)

Figure 52 *Caryedon serratus*, adult, showing antennae and enlarged femur of hind leg

Figure 53 *Caryedon serratus*, adult, hind leg showing enlarged femur and row of small spines on ventral surface

Zabrotes (Figures 54–55)

- Tibia of hind leg with two long movable spurs at tip (Figure 55)

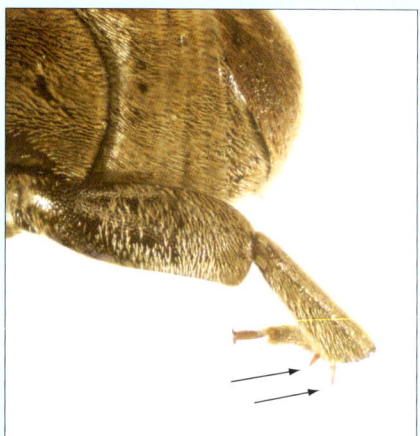

Figure 54 *Zabrotes subfasciatus*, adult, showing pattern on elytra

Figure 55 *Zabrotes subfasciatus*, adult, hind leg, spines at tip of tibia on hind leg

Identification of *Acanthoscelides* species collected in rural environments in tropical America needs to be undertaken with care. More than 250 species are known and a few other than *A. obtectus* have occasionally been associated with stored products, e.g. *A. zeteki*. Outside this area it is unlikely that species other than *A. obtectus* infest stored beans.

Life cycle

Eggs are laid individually on the seed coat or pod. Eggs of *Acanthoscelides* spp. are laid loose but those of the other genera are glued firmly to the seed or pod. Newly hatched larvae burrow directly into the seed. As they do so, the case of an attached egg becomes white (and obvious to see, especially in dark seeds) as the egg shell is filled with frass from the excavations. Larvae complete development within an individual seed in which they excavate a large cavity. Before pupation, larvae eat a round hole to the surface of the seed but leave the seed coat intact. This is visible on the seed surface as a circular translucent 'window'. Pupation takes place within the cavity in the seed; the adult emerges by biting and pushing its way through the 'window', leaving a neat circular hole in the seed (Figures 38, 44). Adults do not feed on the dried seed but may consume nectar and pollen from flowers. Adult lifespan is very variable. It may be less than 10 days for populations confined to stored pulses stored under tropical conditions but may be greater than 100 days when adults hibernate and/or have access to flowers and nectar. Populations of *Bruchus* and *Bruchidius* spp. survive as adult beetles between crops by hiding under bark or in crop debris.

Physical limits and optimum rate of multiplication

Species	Conditions within which breeding takes place	Shortest development period, with optimum conditions	Maximum monthly rate of increase
Acanthoscelides obtectus	15–33°C	27 days at 30°C, 80% r.h.	25
Bruchus pisorum	c. 60 days		
Callosobruchus analis	18.5–37°C	27 days at 32.5°C, 90% r.h	
Callosobruchus chinensis	17.5–37°C	22 days at 32°C, 90% r.h	
Callosobruchus maculatus	18–37°C, 20–90% r.h.	21 days at 32°C, 90% r.h.	50
Callosobruchus phaseoli	13–35°C		
Callosobruchus rhodesianus	17–35°C	25 days at 30°C, 70% r.h.	
Caryedon serratus	23–35°C	42 days at 30–33°C, 70–90% r.h.	
Zabrotes subfasciatus	20–38°C	24–27 days at 32°C, 70% r.h.	

The number of generations of *Bruchus* and *Bruchidius* that occur in a year is linked to the cropping cycle and the availability of ripening seeds. Often this amounts to only one generation a year. Development of populations of bruchids capable of infesting pulses in storage is rapid and continuous, especially under tropical conditions. Compared to most other storage pests, population development of *C. chinensis* and *C. maculatus* is especially rapid. *A. obtectus* and *C. phaseoli* can both breed at relatively low temperatures that allows them to become significant pests in more temperate regions.

Economic importance

In terms of the damage they cause, *A. obtectus*, *Callosobruchus* spp. and *Z. subfasciatus* rank among the most important insect pests of stored products. These pests cause considerable economic damage especially under conditions of tropical subsistence agriculture. A combination of factors make this so. These include climatic conditions for optimal population growth, use of small storage structures (e.g. bags and baskets) that allow easy access by the beetle to the stored commodity, close proximity of production and storage places – making pre-harvest infestation likely, and the limited availability and use of chemical control measures. Even in commercial storage, infestations can be severe, especially in warm to hot climates and when commodities are stored in bags. Damage to bulk-stored pulses is likely to be less severe and be limited to surface layers as bruchids cannot easily penetrate deep into bulks, especially of small-seeded commodities.

Bruchid species are quite specific as to which pulse they attack. Species that attack beans (*Phaseolus* spp.) generally do not attack pulses of the genus *Vigna* and *vice versa*. This is an important aid to the identification of the genus involved (see below).

Host range of bruchid pests of stored pulses

Species	Peanut, groundnut *Arachis*	Chickpea *Cicer*	Pigeon pea *Cajanus*	Soybean *Glycine*	Lentils *Lens*	Dolichos beans *Lablab*	Garden pea *Pisum*	Beans *Phaseolus*	Broad bean *Vicia*	Mung beans, grams, cowpeas *Vigna*	Bambarra groundnuts *Voandzeia*
Acanthoscelides obtectus								X	X		X
Acanthoscelides zeteki			X								
Bruchus spp.					X		X		X		
Bruchidius spp.					X				X	X	
Callosobruchus analis				X						X	
Callosobruchus chinensis		X			X	X?	X			X	
Callosobruchus maculatus		X	X	X	X	X?	X			X	
Callosobruchus phaseoli					X	X	X			X	
Callosobruchus rhodesianus							X			X	
Callosobruchus subinnotatus											X
Callosobruchus theobromae	X		X	X							
Caryedon serratus ▼	X										
Zabrotes subfasciatus									X	X#	

? Status unclear
Some strains capable of attacking this commodity.
▼In addition to attacking groundnuts, *C. serratus* is also a pest of Tamarind – *Tamarindus indica* L. (Caesalpinioidiae)

Type of damage and symptoms

Damage is distinctive (Figures 38, 45). As adults emerge from seeds they leave behind neat circular holes in the seed, behind which is a large cavity left by the larvae. Loss of seed material is considerable – each adult *Callosobruchus* emerging from a cowpea (*Vigna unguiculata*) would have consumed about 25% of the seed from which it emerged. Damaged seed often does not germinate or germinate well. Heavy infestations of bruchids can cause heating of commodity which results in quality loss and mould growth.

Ecology

Bruchids that attack pulse crops can be divided into two groups. While all genera can attack the ripening crop in the field, members of the first group (*Acanthoscelides, Callosobruchus, Caryedon* and *Zabrotes*) are able to continue breeding on dried pulses in storage. Members of the second group (*Bruchus* and *Bruchidius* spp.), are important field pests of ripening pulses but are not able to infest dried stored seed. Infestations of these genera do not persist in storage.

Where storage sites are close to areas where crops are grown, adults beetles may move from heavily infested store residues into the field to infest crops as they mature. Even if the store is carefully cleaned out the insect may return in field-infested seeds. Populations of some pest species may also persist in wild legume populations. For example, in Japan, *C. chinensis* is known to infest the wild legume *Vigna angularis* var. *nipponensis*.

Bruchid beetles have closely co-evolved with their host plant species, having overcome the antifeedant chemicals within these seeds that prevent attack by most other storage pests. Research is under way to manipulate the presence and quantity of these substances to confer resistance. In general however, the development of large-seeded varieties that are attractive and palatable to humans and animals has inadvertently led to a reduction in the quantity of antifeedant chemicals present and has lead to an increased susceptibility of many modern varieties to bruchid pests.

Given time and sufficient evolutionary pressure bruchids are able to extend their capability to attack new species of pulses – as witnessed by the relatively recent development of stains of *Z. subfasciatus* capable of breeding on *Vigna* spp. Genera such as *Acanthoscelides* and *Callosobruchus* contain large numbers of species, many of which attack wild relations of crop plants and some of these over time could develop into storage pests.

Monitoring

Infestations quickly become obvious as a result of the presence of eggs stuck to the outside of seed together with the presence of the active adults and emergence holes in seeds.

Acoustic detectors are commercially available which detect the sounds that internal stages of these beetles make during feeding. X-ray photography has been used to detect hidden larvae developing within grains. Traditionally, prior to consumption many consumers have used flotation in water to separate infested from uninfested grains. Infested grains typically float whereas intact grains mostly sink.

Geographical distribution

Species	Pest status	USA & Canada	Central & South America	Europe & N.Asia	Mediterranean basin	Africa	S. & SE. Asia	Australia & Oceania
Acanthoscelides obtectus	●●●●	X	X	X®	X	X	X®	X
Acanthoscelides zeteki	●●		X®					
Bruchus spp.	●●	X	X	X	X		X	X
Bruchidius spp.	●●	X		X	X	X	X	X
Callosobruchus analis	●●●●						X	
Callosobruchus chinensis	●●●●	X			X	X	X	X

Species	Pest status	USA & Canada	Central & South America	Europe & N.Asia	Mediterranean basin	Africa	S. & SE. Asia	Australia & Oceania
Callosobruchus maculatus	●●●●	X	X	X	X	X	X	X
Callosobruchus phaseoli	●●●		X	X	X	X	X	X
Callosobruchus rhodesianus	●●●					X		
Callosobruchus subinnotatus	●●					X®		
Callosobruchus theobromae	●●					X	X	
Caryedon serratus	●●		X		X®	X	X	X
Zabrotes subfasciatus	●●●●	X	X		X	X	X	

Pest status: ● minor to ●●●● major pest
X: recorded
®: restricted distribution

Species of *Callosbruchus*, *Caryedon* and *Zabrotes* are most often encountered in warm temperate to tropical climates, especially in Africa and Asia. Some *Callosobruchus* species are still relatively restricted in distribution and therefore have potential for further spread. *C. rhodesianus* largely replaces *C. chinensis* in southern Africa. *A. obtectus* is more often encountered in temperate climates than the other major genera mentioned above, it is widespread in the tropics but is largely absent from Asia; however, it has been reported from the Philippines.

Bruchus and *Bruchidius* species are most often encountered in temperate, Mediterranean and hot dry regions, such as areas of south-western Asia.

References
Aitken (1975), Arbogast (1991), Haines (1989, 1991) Kingsolver (1987), Southgate (1979).

Weevils
(Family: Curculionidae)

Caulophilus oryzae	Broad-nosed weevil
Sitona spp.	Sitona weevils
Sitophilus granarius	Granary weevil
Sitophilus linearis	Tamarind weevil
Sitophilus oryzae	Rice weevil
Sitophilus zeamais	Maize weevil

Summary

Feeding strategy	primary pest
Commodities attacked (major pest species)	whole cereal grains, solid cereal products, some pulses
Distribution	worldwide
Economic importance	grain feeding *Sitophilus* spp. – high
Eggs	inserted individually into grain
Larvae	apodous (legless), immobile, live concealed within grain
Adults	long lived, feed on commodity, can fly – except *S. granarius* which is flightless

Introduction

The Curculionidae or true weevils is the largest family of beetles known. They are found in a wide range of habitats and many species are important pests of agriculture, horticulture and forestry. They attack the stems, roots and seeds of plants and some are wood borers. The head of many adult weevils has a characteristic snout. Members of the genus *Sitophilus* are among the most important pests of stored grain. Other genera are minor pests or are accidental on grain.

Identification

The distinctive appearance of these insects (Figures 56–67) makes it unlikely that they will be confused with other common beetle pests of stored products. *Sitophilus* spp. are brown to black insects, 2.5–4 mm long (Figure 57–67). The size of adult varies somewhat with seed size from which it emerged. Grain feeding species can be identified to species as below. Externally *S. zeamais* and *S. oryzae* are identical; however, they may be reliably separated by examination of genitalia which is easiest to perform with male specimens. Adults are easy to sex by external characteristics.

S. linearis (Figure 60) is usually identifiable by its close association with tamarind (*Tamarindus indica*) pods. Other genera of weevils are sometimes found on stored grain. Sitona weevils (*Sitona* spp.) are pests of clover often grown under wheat crops and can be a contaminant of harvested grain in many temperate regions (Figure 56). At 6 mm long, they are larger than *Sitophilus* spp. and lack a long snout. *Caulophilus oryzae* is similar in appearance to *Sitophilus*, but has a shorter fatter snout and lacks the spots on the elytra.

Larvae of weevils are legless (apodous) (Figure 65).

Far left: Figure 56 *Sitona* spp., adult, showing elbowed antennae, lack of snout, markings on prothorax

Left: Figure 57 *Sitophilus granarius*, adult, showing long snout, oval pits on thorax, heavily ridged elytra without coloured spots

Figure 58 *Sitophilus granarius*, adult, thorax and elytra, oval pits on thorax and heavily ridged elytra

Figure 59 *Sitophilus granarius*, adults, live on wheat

Figure 60 *Sitophilus linearis*, adult

Figure 61 *Sitophilus oryzae*, adult, showing round pits on thorax, coloured patches on elytra

0.2 mm

Figure 62 *Sitophilus oryzae*, adult, aedeagus, convex outer surface

Figure 63 *Sitophilus oryzae*, male, showing snout short relative to width with heavy irregular pitting

Figure 64 *Sitophilus oryzae*, female, showing snout long relative to width with light regular pitting

Figure 65 *Sitophilus oryzae*, larva, without legs

Figure 66 *Sitophilus oryzae*, X-ray of internal infestation of wheat

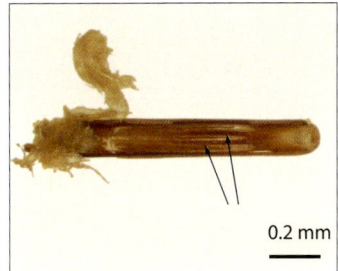

0.2 mm

Figure 67 *Sitophilus zeamais*, adult, aedeagus, outer surface with three longitudinal ridges

Key to grain-feeding *Sitophilus* species

1   Colour – uniform dark brown, elytra not marked with spots, flight wings (under elytra) absent, pronotum marked with oval shaped punctures (Figure 57–59) . . . .
. . . . *Sitophilus granarius*
Colour dark to black in mature specimens, each elytra marked with two dull orange or yellow spots. Flight wings (under elytra) present, pronotum marked with circular-shaped punctures (Figure 61)   2

2   (By dissection) outer surface of aedeagus (functions as penis) of male smooth and convex in cross-section (Figure 62)   . . . . *Sitophilus oryzae*
(By dissection) outer surface of aedeagus of male with two grooves in cross-section (Figure 67)   . . . . *Sitophilus zeamais*

To sex adult *Sitophilus*

Rostrum (snout) is relatively short and wide, punctures along rostrum irregular and large, often touching each other to give 'rough' appearance (Figure 63) . . . . male
Rostrum (snout) is relatively long and narrow, punctures along rostrum in regular rows and small and not touching each other to give 'smooth' appearance (Figure 64) . . . female

Life cycle

Eggs are laid singly into grains. The female selects a spot on the grain surface then chews a small hole, into which she lays an egg. The hole is then plugged with a waxy secretion. *Sitophilus* females lay up to 150 eggs over their lifespan. Larvae develop hidden within a cavity hollowed out within the grain. Larvae are cannibalistic and larger larvae may eat less developed individuals should they meet. Pupation takes place within the cavity made by the larva and upon emergence from the pupa the adult may spend several days within the cavity. Eventually it will chew its way out, leaving a ragged hole. Adult beetles continue to feed on grain and are long lived (lifespan of 3 to 6 months, longer in cooler months).

Physical limits and optimum rate of multiplication

Species	Conditions within which breeding takes place	Shortest development period, with optimum conditions	Maximum monthly rate of increase
Sitophilus granarius	11–34°C, > 40% r.h.	25 days at 30°C, 70% r.h.	15
Sitophilus oryzae	15–34°C, > 40% r.h.	25 days at 30°C, 70% r.h	25
Sitophilus zeamais	15–34°C, > 40% r.h.	25 days at 30°C, 70% r.h	25

S. oryzae and *S. zeamais* breed most rapidly under humid tropical conditions. *S. granarius* is cold-hardy, and the low minimum temperature at which this species breeds allows it to be an important pest in temperate regions.

Economic importance

Caulophilus oryzae is a minor pest of maize most frequently encountered under conditions of tropical subsistence agriculture, where infestation can begin prior to harvest.

Sitophilus spp. are major pests of whole cereal grains and dried root crops such as cassava. Along with *Rhyzopertha dominica* and *Prostephanus truncatus* (Col. Bostrichidae), they rank as the most important pests of whole grain. *Sitophilus oryzae* is a well known pest of dried pasta. Certain strains of *S. oryzae* can attack stored pulses, including chickpeas (*Cicer arietinum*), cowpeas (*Vigna unguiculata*) and peas (*Pisum sativum*).

If left unchecked, infestations of *Sitophilus* spp. can cause devastating damage to stored grain. Currently, in well-run bulk grain storage systems, *Sitophilus* infestations are generally kept under control, but the potential threat remains. They remain a serious pest of farm-stored grain and of bag-stored grain in warehouses, especially under tropical conditions.

S. linearis can cause severe damage to tamarind pods when stored under tropical conditions.

Type of damage and symptoms

Damage is distinctive (Figure 66). Feeding by larvae leaves large cavities inside grains and newly emerging adults leave behind large ragged emergence holes. Adults cause further damage by feeding, mainly by attacking previously damaged grain. *Sitophilus* infestations produce a lot of heat and moisture – this encourages extensive quality loss, mould growth and growth of populations of other insect species.

Ecology

Prior to the 1960s, *S. oryzae* and *S. zeamais* were thought to be one species. It is now understood that they are two very closely related species with somewhat different behavioural characteristics and commodity preferences. For example, *S. zeamais* is more likely than *S. oryzae* to infest standing crops immediately prior to harvest. *S. zeamais* is usually the dominant *Sitophilus* in subsistence agricultural systems whereas *S. oryzae* tends to be more common in warehouse and commercial storage systems. *S. zeamais* is more commonly associated with maize and rice and *S. oryzae* is more often found on wheat, barley and processed cereals. Strains of *S. oryzae* are adapting to feed on pulses and there is a danger that some of these may become significant pests on these commodities.

Under tropical conditions, *S. granarius* is out-competed by either *S. oryzae* or *S. zeamais*. However, *S. granarius* is very cold-hardy and, as a result, is the dominant species in temperate regions. Being flightless, it is largely reliant on human activity for dispersal. *S. granarius* has a long association with stored products and has been a pest since the time of the ancient Egyptians.

Within grain bulks *Sitophilus* populations often aggregate, making them difficult to detect. However, when such aggregations run out of food, adults will disperse, at which stage the infestation may quickly become visible. By this stage, significant damage to the grain mass is likely to have occurred.

Sitophilus spp. often infest grain along with the bostrichids or *Trogoderma granarium* (Col.: Dermestidae). *Sitophilus* tends to dominate under warm humid conditions, such as those in the humid tropics. In hot dry tropical regions, one of the other species tends to dominate.

Monitoring

Sitophilus spp. can be trapped with a range of commercially available pitfall and probe traps placed at the surface or inserted into grain bulks. Efficacy of traps may be improved by the addition of baits such as cracked grain and grain oils and synthetic aggregation pheromone. Adults can be detected by sieving of grain. In addition, adult *Sitophilus* will rush to the surface of sample of grain that is shaken or tapped firmly.

Acoustic detectors are commercially available which detect the sounds that internal stages of these beetles make during feeding. X-ray photography has been used to detect hidden larvae developing within grains.

Geographical distribution

Species	Pest status	USA & Canada	Central & South America	Europe & N.Asia	Mediterranean basin	Africa	S. & SE. Asia	Australia & Oceania
Caulophilus oryzae	●	X®	X					
Sitona spp.	●	X		X	X		X®	X
Sitophilus granarius	●●●	X	X®	X	X	X ®	X®	X
Sitophilus linearis	●●	X®	X		X®	X®	X	X®
Sitophilus oryzae	●●●●	X	X	X	X	X	X	X
Sitophilus zeamais	●●●●	X	X	X	X	X	X	X

Pest status: ● minor to ●●●● major pest
X: recorded
® restricted distribution

Caulophilus oryzae is a Central American species which extends to the West Indies and the far south of the USA. *Sitona* species are widespread in areas with a temperate or Mediterranean climate.

Grain feeding *Sitophilus* are found worldwide, *S. zeamais* is most usually encountered in the tropics and *S. granarius* is largely restricted to temperate regions, including cool highland areas in tropical latitudes. *S. oryzae* is established everywhere except the coolest temperate regions. For example in the Australian wheat-belt, *S. zeamais* is found in northern sub-tropical grain growing areas, *S. oryzae* is widely distributed and *S. granarius* is restricted to southern temperate regions. A similar pattern can be observed in the Americas with *S. granarius* being found in temperate grain growing regions such as Canada and Argentina. *S. oryzae* is widespread in all but the coldest areas and *S. zeamais* dominates in tropical maize growing regions. *S. linearis* is recorded from tropical regions where tamarinds are grown.

References

Aitken (1975), Arbogast (1991), Haines (1991).

Dermestid beetles (Family: Dermestidae)

Introduction to family

The Dermestidae contains about 1000 described species belonging to about 50 genera. In nature, most feed mainly on dried material of animal origin but some, notably *Trogoderma* species, feed wholly or partly on dried plant material. Dermestids are typically found in nests of birds, insects and mammals, on carrion and under bark. Adults of some species are nectar feeders and can be found visiting flowers.

Adults of species associated with stored products measure 2–12 mm in length. They are round or oval in shape and are to a variable extent covered with setae or scales. In some genera, especially *Anthrenus,* these form colourful and distinctive patterns.

Larvae of dermestidae are distinctive in appearance. They are eruciform, elongate or oval in shape and distinctively covered with a range of long and short hairs or setae. Some species have tufts of barbed 'arrow headed' setae known as hastisetae, which are unique to the Dermestidae.

In the storage environment by far the most important genera are *Trogoderma* (pages 66–72) and *Dermestes* (Pages 61–66). The genus *Trogoderma* contains one of the most feared pests of stored grain – the khapra beetle *T. granarium*. *Dermestes* are major pests of hides, skins and dried fish. Other genera listed are mainly scavengers, but may be significant pests in homes and in museums where they attack wool carpets and garments, artefacts and collections of dried animals and plants. Hairs and cast skins of larval dermestids are highly allergenic and in contact with human skin can sometimes cause severe irritation.

Additional genera of Dermestidae are occasionally recorded from stored products. These include the genera *Megatoma*, *Thorictodes* and *Ressa.* For further details, see Peacock (1993).

Major genera of Dermestidae associated with stored products can be identified by the following key. Some of these characters are difficult to see and will require a binocular microscope with high (x 50+) magnification. This key only covers major genera associated with stored products. If a specimen does not fit the relevant key well, it may belong to one of the genera only occasionally associated with stored products, in which case it is best to utilize the more comprehensive keys of Kingsolver (1987) and Peacock (1993).

Simplified key to major genera of adult Dermestidae associated with stored products (based on Kingsolver 1987 and Peacock 1993)

1   Median ocellus (simple eye in middle of forehead) present (Figure 69), adult length usually less than 5 mm 2
     Median ocellus absent, beetles 5–10 mm in length . . . . . *Dermestes* spp. (Figures 80, 89)

2   Body of adult covered in colourful scales (Figure 68), also cavity into which antennae fits fully visible when viewed from the front (Figure 69) . . . . *Anthrenus* spp.
     Body with hairs (setae) not scales, also cavity into which antennae fits not visible or only partly visible when viewed from the front (Figure 74) 3

3   First segment of hind tarsi half or less the length of the second segment (Figure 78). Edge of antennal cavity not well defined (Figure 75) . . . *Attagenus* spp. (Figure 73–79)
     First segment of hind tarsi as longer or longer than second segment (Figure 99). Edge of antennal cavity well defined, surface of cavity smoother than surrounding area of thorax (Figures 96, 102) 4

4   Antennal club with 3 to 8 segments, gradually differentiated from each other and rest of antennae, segments joined symmetrically along central axis (Figure 95) . . . .
. . . .*Trogoderma* spp. (Figures 93–104)
Antennal club with 3 large segments, much bigger than rest of segments of antennae, segments of club joined to one side (Figure 72) . . . .*Anthrenocerus* (Figure 71)

Simplified key to major genera of larvae of Dermestidae associated with stored products (based on Kingsolver 1987 and Peacock 1993)

Identification of cast skins is usually possible

1   Ninth segment of abdomen with two horn-like structures (urogomphi) (Figure 85) . . .
. . . .*Dermestes* spp. (Figure 84)
Ninth segment of abdomen without two horn-like structures (Figure 103) 2

2   Dorsal surface of larvae with hastisetae (barbed arrow-headed setae – Figure 104). These are often partly rubbed off on worn specimens (Figures 70, 103, 104) 3
Dorsal surface of larvae without hastisetae . . . . *Attagenus* spp. (Figure 76)

3   Tufts of hastisetae emerge directly from sclerotized (brown) areas of abdominal segments 4–7, hastisetae do not overlap over middle of abdomen. hastisetae present on final abdominal segment. . . *Trogoderma* spp. (Figure 103–104) and *Anthrenocerus* spp. Tufts of hastisetae emerge from non-sclerotized area below sclerotized (brown) areas of abdominal segments 4–7. Tufts of hastisetae overlap over middle of abdomen (Figure 70), hastisetae not present on final abdominal segment. . . . .*Anthrenus* spp.

References

Aitken (1975), Arbogast (1991), Banks (1994), Kingsolver (1987), Haines (1991), Haines and Rees (1989), Hinton (1945), Mound (1989), Peacock (1993), Roach (2000).

Variegated carpet beetles, Museum beetles
(*Genus: Anthrenus*) (selected species)

Anthrenus flavipes	Furniture carpet beetle
Anthrenus verbasci	Small cabinet beetle, Varied carpet beetle, Variegated carpet beetle

Summary

Feeding strategies	primary pest, secondary pest, scavenger
Commodities attacked	dried material of animal origin
Distribution	worldwide
Economic importance	high as domestic and museum pest
Eggs	laid amongst commodity
Larvae	eruciform, mobile, live amongst commodity
Adults	short lived, do not feed on commodity, can fly

Introduction

The genus *Anthrenus* comprises more than 80 species worldwide. Members of this genus are mainly scavengers on dried material of animal origin. Many species are from time to time found associated with human activities as scavengers and pests of materials of animal origin. By far the most frequently encountered species is *A. verbasci*.

Identification

To identify *Anthrenus* species from other Dermestids see keys above or those of Kingsolver (1987), Haines (1991), Hinton (1945) or Peacock (1993). The two species listed here are often associated with the storage environment. However, many other species are known, some of which are found in domestic situations.

Adult *Anthrenus* are distinctive small oval 'seed-like' beetles 2–3 mm long (Figures 68–69). Legs and antennae can be tucked in when the insect is alarmed. The upper surface of the adult is covered with scales, which in all common pest species are multicoloured (mainly browns, black and white) giving the insect a characteristic patterned mottled appearance. Larvae are oval in shape, creamy white to light brown in colour and are clothed in transverse bands of hairs (Figure 70). Tufts of hairs or hastisetae are present as patches on either sides of abdominal segments and overlap over the middle of abdomen.

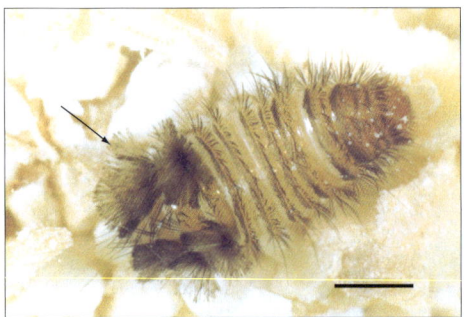

Top left: Figure 68 *Anthrenus verbasci*, adult, showing body covered in highly patterned scales

Top right: Figure 69 *Anthrenus verbasci*, adult, head / antennae showing median ocelus, antenna cavity fully visible from front

Left: Figure 70 *Anthrenus verbasci*, larva, showing tufts of hairs including hastisetae at tip of abdomen

Life cycle

Females lay 30–100 eggs. On hatching, larvae feed and burrow into infested material, moulting many times as they develop. If conditions become unfavourable, larvae may enter diapause (a condition of suspended animation) and in temperate regions they usually overwinter in this state. Pupation takes place inside the skin of the last larval instar. Adults live for 2–4 weeks (*A. verbasci*) and typically emerge in late spring and early summer. They fly well and can often be found feeding on nectar and pollen from flowers. This food is not essential as populations can survive without it.

Physical limits and optimum rate of multiplication

Species	Conditions within which breeding takes place	Shortest development period, with optimum conditions	Maximum monthly rate of increase
Anthrenus flavipes	20–35°C	> 100 days at 30–35°C	

Multiplication is slow. Species that live under ambient conditions in temperate regions can take one or two years to complete development. Several generations a year may be possible under tropical conditions. Larvae can survive long periods of cold and dry conditions.

Economic importance

Anthrenus spp. are important pests of museums and homes, attacking dried artefacts of organic origin, including preserved insect specimens, skins, hides and woollen goods including clothes and carpets. Larvae of *A. verbasci* have occasionally been found feeding on a range of stored foodstuffs including savoury biscuits, dried baby food, cakes, peanuts, wheat and maize. In commercial grain storages, *Anthrenus* are usually scavengers of minor importance being associated with birds' nests or animal remains. In temperate regions, *A. verbasci* is a major household pest of woollen carpets and textiles.

Type of damage and symptoms

Damage consists of holes eaten by larvae in infested material and packing material. Large numbers of cast skins are left scattered through the infested and damaged material.

Ecology

In nature, *Anthrenus* species are scavengers of material mainly of animal origin. Household infestations typically start from bird's nests and dead birds or rodents in roof spaces or wall cavities. Emergence of adults in many species is synchronised to take advantage of optimal ambient conditions and easy availability of flowers for adult food. This sense of timing is very strong and persists even in cultures kept in constant conditions independent of outside stimuli.

Monitoring

Infestations are usually spotted by accumulations of cast larval skins. A synthetic sex pheromone is commercially available to attract adult male beetles. Adults are attracted to light and can be found feeding on pollen and nectar on flowers.

Geographical distribution

Species	Pest status	USA & Canada	Central & South America	Europe & N.Asia	Mediterranean basin	Africa	S. & SE. Asia	Australia & Oceania
Anthrenus flavipes	●●	X	X	X	X	X	X	X
Anthrenus verbasci	●●	X	X	X	X	X	X	X

Pest status: ● minor to ●●●● major pest
X: recorded

Both listed species are found worldwide. *A. verbasci* is common in temperate regions and is frequently found in natural habitats there. *A. flavipes* is more tropical in distribution. In temperate regions, it is largely restricted to heated environments.

Australian carpet beetle
(*Anthrenocerus australis*)

Summary

Feeding strategies	secondary pest, scavenger
Commodities attacked	dried material of animal and vegetable origin
Distribution	Australia, New Zealand, north-west Europe
Economic importance	low–medium
Eggs	laid amongst commodity
Larvae	eruciform, mobile, live amongst commodity
Adults	long lived, do not feed on commodity, can fly

Introduction

The genus *Anthrenocerus* are scavengers on dried material of animal origin. This genus originates in Australia where more than 30 species are described. *A. australis* has become associated with human activities as a scavenger and pest of materials of animal origin. It has been spread by trade to New Zealand and Europe.

Identification

To identify *Anthrenocerus australis* from other Dermestidae see key above or Peacock (1993). Roach (2000) describes numerous *Anthrenocerus* species present in Australia.

Anthrenocerus australis is an oval, hairy beetle 2.2–2.5 mm long (Figures 71–72). Background colouring is mid- to dark brown. The pronotum is marked with patches of light-coloured setae and three wavy bands of light-coloured setae run across the elytra.

Larvae are eruciform, hairy and elongate, and when full grown they are about 7 mm long.

Figure 71 *Anthernocerus australis*, adult showing patterning on prothorax and elytra

Figure 72 *Anthernocerus australis*, adult, head / antennae, antennal club with asymmetrically jointed segments

Life cycle

Eggs are laid in cracks and crevices in the infested material. Larvae feed and burrow into infested material, moulting many times.

Physical limits and optimum rate of multiplication

Multiplication is slow. If feeding on dried insects, development may take up to a year and longer when feeding on wool.

Economic importance

A. australis is a pest of museums and domestic situations attacking dried artefacts of organic origin and woollen goods. In parts of Australia, it is an important domestic pest of wool carpets. It is occasionally found in grain storages, mainly infesting residues.

Type of damage and symptoms

Damage usually consists of holes eaten by larvae in infested material and packing material. Large numbers of cast skins, which remain intact, become scattered through the infested and damaged material.

Ecology

In nature, *A. australis* are scavengers of material mainly of animal origin. Infestations typically start from bird's nests and dead birds or rodents in roof spaces or wall cavities. They may also be found in the nests of bees, ants and wasps. Adults can also be found feeding on nectar on flowers.

Monitoring

No specific information available. Infestations are usually spotted by accumulations of cast larval skins.

Geographical distribution

Species	Pest status	USA & Canada	Central & South America	Europe & N.Asia	Mediterranean basin	Africa	S. & SE. Asia	Australia & Oceania
Anthrenocerus australis	●●			X®				X

Pest status: ● minor to ●●●● major pest
X: recorded
®: restricted distribution

A. australis is of Australian origin, where it is widespread in southern parts of the country. It is also found in New Zealand and has been introduced into north-west Europe, where it has been recorded from Belgium, Netherlands, Germany and the UK.

Black carpet beetles, Fur beetles
(Genus: *Attagenus*)

Attagenus brunneus	
Attagenus cyphonoides	
Attagenus fasciatus	Tobacco seed beetle
Attagenus megatoma	Black carpet beetle
Attagenus pellio	Two spotted carpet beetle, Fur beetle
Attagenus smirnovi	
Attagenus unicolor	Black carpet beetle

Summary

Feeding strategies	secondary pest, scavenger
Commodities attacked	dried material of animal and vegetable origin
Distribution	worldwide
Economic importance	important household and museum pest
Eggs	laid amongst commodity
Larvae	eruciform, mobile, live amongst commodity
Adults	long lived, do not feed on commodity, can fly

Introduction

The genus *Attagenus* are mainly scavengers on dried material of animal origin. About 130 species are known worldwide of which less than ten are frequently associated with human activities as scavengers and pests of materials of animal origin.

Identification

To identify *Attagenus* species from other Dermestidae, see keys above or Kingsolver (1987), Haines (1991) or Peacock (1993). For further information on species of this genus, see Hinton (1945) or Peacock (1993). *Attagenus* are oval, hairy beetles 3–5 mm long (Figures 73–79). Several species associated with stored products are uniform black in colour (Figure 79). Two species are distinctive. *A. fasciatus* has dark wing cases with a distinctive pale band running across them, about half way across the wing case (Figure 73). *A. pellio* has black elytra each with a small white spot in the centre (Figure 77).

Larvae are eruciform, hairy and elongate, and when full grown are about 10 mm long (Figure 76).

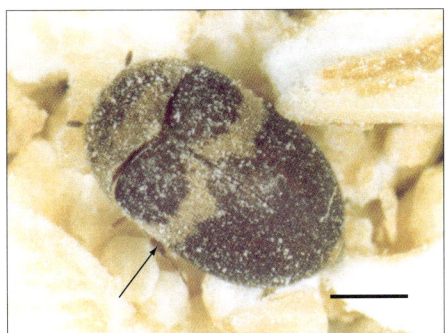

Figure 73 *Attagenus fasciatus*, adult, live showing pale band on elytra

Figure 74 *Attagenus fasciatus*, adult, head, front, antenna cavity not fully visible from front

Figure 75 *Attagenus fasciatus*, adult, head from below showing antennae and antennal cavity

Figure 76 *Attagenus fasciatus*, larva

Figure 77 *Attagenus pellio*, adult, showing white spot on each elytra

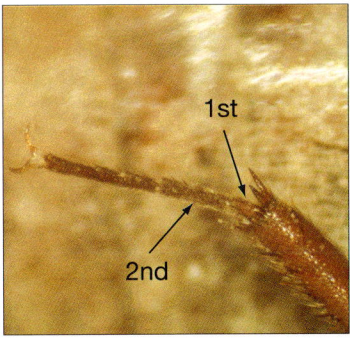

Figure 78 *Attagenus* spp., adult, tarsus of hind leg, first segment less than half length of second

Figure 79 *Attagenus unicolor*, adult, showing uniform black elytra

Life cycle

Eggs are laid in cracks and crevices in the infested material. Larvae feed and burrow into infested material moulting many times. In temperate regions, *Attagenus* species pass the winter either as adults or as larvae. Pupation takes place within the skin of the last larval instar. Adults fly very well and can be found feeding on nectar and pollen from flowers during early summer. This is not essential for survival as populations can survive in laboratory culture without access to such food.

Physical limits and optimum rate of multiplication

Species	Conditions within which breeding takes place	Shortest development period, with optimum conditions	Maximum monthly rate of increase
Attagenus pellio	15–30°C	6 months	
Attagenus smirnovi		113 days at 24°C, 70–80% r.h	
Attagenus unicolor		c. 240 days at 24°C, 70% r.h	

Multiplication of *Attagenus* is slow. Even under optimal conditions, the life cycle may take six months to complete, but in temperate climates, development to adulthood may take up to 2–3 years. Larvae can withstand very cold and dry conditions and recommence growth when temperature is suitable.

Economic importance

Like *Anthrenus* spp., *Attagenus* are pests in museums and domestic situations where they attack dried artefacts of organic origin including skins, hides and woollen goods, oilseeds, fish meal and dog biscuits. In grain stores, *Attagenus* spp. are minor pests or scavengers. In Japan, *A. unicolor* has been reported as a pest of silkworm pupae and cocoons.

Type of damage and symptoms

Damage usually consists of holes eaten by larvae in infested material and packing material. Large numbers of cast skins, which often remain intact, become scattered through the infested and damaged material.

Ecology

In nature, *Attagenus* species are scavengers of material mainly of animal origin. Household infestations typically start from bird's nests and dead birds or rodents in roof spaces or wall cavities. They may also be found in the nests of bees and wasps. Adults can also be found feeding on nectar on flowers.

Monitoring

Infestations are usually spotted by accumulations of cast larval skins. A synthetic sex pheromone is commercially available to attract adult male beetles. Adults are attracted to light and can be found feeding on pollen and nectar on flowers.

Geographical distribution

Species	Pest status	USA & Canada	Central & South America	Europe & N.Asia	Mediterranean basin	Africa	S. & SE. Asia	Australia & Oceania
Attagenus brunneus	●●	X		X	X		X	
Attagenus cyphonoides	●●	X	X	X	X	X	X	
Attagenus fasciatus	●●		X	X	X	X	X	X
Attagenus megatoma	●●	X	X	X	X		X	X
Attagenus pellio	●●	X	X	X	X	X	X	X
Attagenus smirnovi	●●			X		X		
Attagenus unicolor	●●	X		X			X	

Pest status: ● minor to ●●●● major pest
X: recorded

Attagenus spp. occur worldwide in both tropical and temperate regions. In warmer climates, *A. fasciatus* is most often encountered. *A. unicolor* is especially well known as a domestic pest in North America. It is largely replaced in Europe by *A. pellio*. In Russia and parts of Europe and Africa *A. smirnovi* is becoming more frequently encountered as a domestic pest. *A. brunneus* is widely distributed in the northern hemisphere. In Australia, the most frequently encountered species is *A. megatoma*.

Hide beetles, Larder beetles (Genus: *Dermestes*)

Dermestes ater	Black larder beetle
Dermestes carnivorus	
Dermestes frishii	Hide beetle
Dermestes haemorrhoidalis	
Dermestes lardarius	Larder beetle
Dermestes maculatus	Hide beetle
Dermestes peruvianus	Peruvian larder beetle

Summary

Feeding strategies	primary pest, secondary pest, scavenger
Commodities attacked	dried material of animal origin, copra
Distribution	worldwide
Economic importance	medium–high
Eggs	laid amongst commodity
Larvae	eruciform, mobile, live amongst commodity
Adults	long lived, feed on commodity, can fly

Introduction

Members of the genus *Dermestes* are mainly carrion feeders but may also be found associated with bird and animal nests. More than 70 species are described worldwide of which about seven are frequently encountered attacking stored hides, skins and dried fish.

Identification

To identify *Dermestes* from other genera of Dermestidae beetles, see keys above or those of Kingsolver (1987), Haines (1991) or Peacock (1993). For further information see Haines and Rees (1989), Hinton (1945), Mound (1989) and Peacock (1993).

Adult *Dermestes* are elongate oval beetles, slightly flattened, 5.5 to 10 mm long (Figures 80–92). In most species the upper surface is black. The underside of the thorax and elytra are densely covered with either white hairs with black spots at the sides (Figure 86) or brown/golden hair (Figure 87). *Dermestes* spp. could be confused with *Alphitobius* species (Tenebrionidae) and *Tenebroides mauritanicus* (Trogossitidae). Members of these genera are glossy black but do not have undersides covered in hair.

Larvae are eruciform – elongate and covered with spines, when full grown they may be up to 20 mm long (Figure 84). They can be distinguished from other dermestid larvae by having two horn-like structures (urogomphi) on the last abdominal segment (Figure 85). Other dermestid larvae commonly associated with stored products lack these 'horns'.

Figure 80 *Dermestes ater*, adult

Figure 81 *Dermestes ater*, adult, underside covered with brown hairs with patches of darker hairs

Figure 82 *Dermestes carnivorus*, adult, tip of underside, covered with white hairs, with black patches at margin of each segment, no black patch at tip of final segment

Figure 83 *Dermestes frishii*, adult, tip of elytra with no spine

Figure 84 *Dermestes frishii*, larva

Figure 85 *Dermestes frishii*, larva, tip of abdomen, ninth segment with horn-like structures (urogomphi)

Figure 86 *Dermestes frishii*, adult, underside, covered with white setae, with black patches at margin of each segment, black patch at tip of final segment

Figure 87 *Dermestes haemorrhoidalis*, adult, underside covered in light brown hairs

Figure 88 *Dermestes haemorrhoidalis*, adult, elytra, with fringe of hair which projects beyond margin

Figure 89 *Dermestes lardarius*, adult showing pale band across elytra

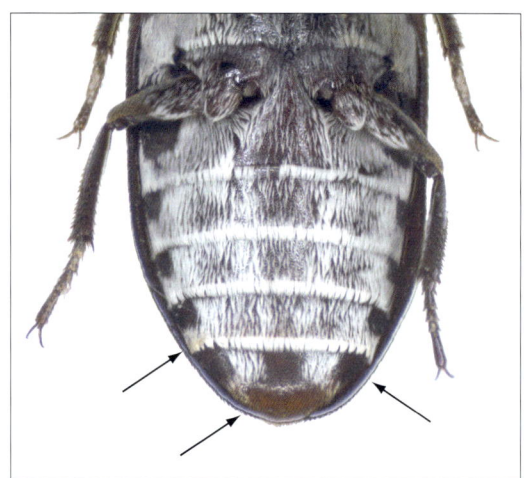

Top left: Figure 90 *Dermestes maculatus*, adult

Top right: Figure 91 *Dermestes maculatus*, adult, underside covered with white setae, with black patches at margin of each segment, black patch at tip of final segment

Right: Figure 92 *Dermestes maculatus*, adult, tip of elytra with spine

Key to adult *Dermestes* species associated with stored products

1 Basal half of elytra with band of pale brown hairs, in the middle of which are small patches of black hairs. Other half of elytra black (Figure 89) . . . . *D. lardarius*
Elytra black, without pale markings (Figures 80, 90) 2

2 Underside of abdomen covered with white hairs and dark spots (Figures 86, 91) 3
Underside of abdomen covered with brown or golden hairs (Figure 81, 87) 5

3 Tip of elytra with spine (Figure 92) . . . . *D. maculatus*
Tip of elytra without spine (Figure 83) 4

4 Tip of underside of abdomen marked with black patch (Figure 86) . . . . *D. frischii*
Tip of underside of abdomen not marked with black patch (Figure 82) . . *D. carnivorus*

5 Underside of abdomen golden with dark patches (Figure 81) . . . . *D. ater*
Underside of abdomen uniformly coloured (Figure 87) 6

6 At apex elytra with thick fringe of hair which project beyond its edge (Figure 88) . . . .
. . . . *D. haemorrhoidalis*
At apex elytra without thick fringe of hair which project beyond its edge . . . .
. . . . *D. peruvianus*

Life cycle

Eggs are laid at random on the foodstuff. Larvae burrow into the food, moulting five to seven times as they develop. Cast skins are left behind in the infested food. These maintain their shape and are often confused with live larvae. Prior to pupation, larvae wander and burrow into wood, plaster and even through soft metal such as lead, to make a chamber in which to pupate. Adults feed on the infested commodity, are long-lived and fly readily.

Physical limits and optimum rate of multiplication

Species	Conditions within which breeding takes place	Shortest development period, with optimum conditions	Maximum monthly rate of increase
Dermestes ater	r.h. > 40%	42 days at 27–30°C, 75% r.h.	
Dermestes carnivorous	20–35°C, 30–90% r.h.	35°C, 90% r.h.	
Dermestes frishii	20–35°C, r.h. > 30%	26 days at 35°C, 90% r.h.	30
Dermestes haemorrhoidalis	15–32.5°C, r.h. > 40%	27–30°C, 75% r.h.	
Dermestes lardarius	15–30°C, r.h. > 40%	18–20°C, 80% r.h.	
Dermestes maculatus	20–40°C, r.h. > 30%	21 days at 30°C, 75% r.h.	30
Dermestes peruvianus	15–30°C, r.h. > 40%	25°C, 80% r.h.	

D. lardarius, D. haemorrhoidalis and D. peruvianus are frequently encountered in temperate countries and can breed at slightly lower temperatures than other species. D. maculatus and D. frishii have similar requirements, except that D. frishii is more cold-hardy than D. maculatus. D. ater and D. carnivorus require a slightly higher minimum relative humidity and are most often found in humid tropical regions.

Development can be rapid (less than four weeks) under ideal conditions. However, under adverse conditions, development can take up to several years. The number of eggs laid is greatly increased if liquid water is available to female beetles.

Economic importance

Dermestes species will feed on almost any material of animal origin or on commodities of plant origin which have a high protein content. They are important pests of uncured skins and hides and dried fish and fish meal. D. maculatus and D. frishii are the species most usually encountered on dried fish. D. ater is frequently found infesting copra but will also attack dried fish. D. carnivorus is known as a pest of dried fish in hot humid areas of south and south east Asia. D. lardarius is a minor pest of domestic premises. D. haemorrhoidalis and D. peruvianus are inhabitants of food processing plants in North America and Europe. Dermestes spp. can cause a lot of damage to structures holding or containing the infested commodity.

In grain stores Dermestes spp. are scavengers, feeding on dead insects, rodents and birds.

Type of damage and symptoms

Both larvae and adults of Dermestes damage skins, hides and dried fish by burrowing into them. Dried fish can be fragmented by heavy feeding activity. Infested commodities become contaminated with insect bodies and cast skins. When mature, larvae will bore into wood or soft plaster to make a chamber in which they pupate. Over time, storage structures, drying racks and even buildings can become severely weakened by this activity.

Ecology

In nature, *Dermestes* spp. assist in the destruction of animal remains. They can be found in birds' nests and are attracted to corpses of dead animals.

Monitoring

Infestations are usually easy to see, given the size of the insect and the accumulations of cast larval skins. Adult beetles may be attracted to food baits and light.

Geographical distribution

Species	Pest status	USA & Canada	Central & South America	Europe & N.Asia	Mediterranean basin	Africa	S. & SE. Asia	Australia & Oceania
Dermestes ater	●●●	X	X	X	X	X	X	X
Dermestes carnivorus	●●	X	X	X	X	X	X	
Dermestes frishii	●●●	X	X	X	X	X	X	X
Dermestes haemorrhoidalis	●●	X	X	X	X	X		X
Dermestes lardarius	●●	X	X	X	X	X	X	X
Dermestes maculatus	●●●	X	X	X	X	X	X	X
Dermestes peruvianus	●●	X	X	X	X			

Pest status: ● minor to ●●●● major pest
X: recorded

Dermestes species associated with stored products have a cosmopolitan or wide distribution. *D. ater* and *D. carnivorus* are mainly tropical, and records in temperate regions are mostly as a result of interceptions in imports. The other species occur in both temperate and tropical regions.

Khapra beetle, Warehouse beetle
(Genus: *Trogoderma*) (selected species)

Trogoderma glabrum	Glabrous carpet beetle
Trogoderma granarium	Khapra beetle
Trogoderma inclusum	Larger cabinet beetle, Mottled dermestid
Trogoderma ornatum	Ornate carpet beetle
Trogoderma sternale	
Trogoderma variabile	Warehouse beetle

Summary

Feeding strategies	primary pest, secondary pest
Commodities attacked	dried material of plant and animal origin
Distribution	worldwide, but individual species have restricted distributions
Economic importance	high
Eggs	laid amongst commodity
Larvae	eruciform, mobile, live amongst commodity
Adults	short lived, do not feed on commodity, some species can fly

Introduction

Member of the genus *Trogoderma* are associated with dried material of animal and sometimes plant origin. The number of species of this genus remains unknown as many species remain undescribed. Some six species are frequently associated with stored material of animal and plant origin. Of those, the khapra beetle, *T. granarium*, ranks among the most feared pests of stored products.

Identification

To identify *Trogoderma* from other Dermestids, see keys above or Kingsolver (1987), Haines (1991) or Peacock (1993). In addition to the species listed here, other *Trogoderma* are occasionally found in the storage environment and many others are recorded from natural habitats.

Adult *Trogoderma* are oval, light to dark brown beetles covered with fine hairs, 1.8 to 3 mm long (females are bigger than males) (Figures 93–102). The elytra are marked with irregular paler markings, which vary in intensity between species and specimens. Larvae are oval in shape and vary in colour from whitish yellow when young to reddish brown when mature and are clothed in transverse bands of hairs (Figure 103). Tufts of short barbed hairs known as hastisetae (Figure 104) are found on either sides of the final abdominal segments but do not overlap over the middle of the abdomen.

Correct identification of *Trogoderma* found in stored products is difficult. Identification of both adult and larvae to species requires skill and training, a microscope plus an appropriate identification key such as Banks (1994), Hinton (1945), Kingsolver (1987) and Peacock (1993). Adults of *T. granarium* can sometimes be distinguished from other important pest species by its light brown colour and general absence of markings on the elytra (Figure 94). The elytra of other species are usually obviously patterned. Any population of *Trogoderma* without elytral patterns doing damage to stored products should be especially closely examined, just in case they are *T. granarium*. Adults of *T. inclusum* are distinctive, having an obvious notch in the inner margin of the eye (Figure 98). *T. variabile* is also relatively distinct, as elytra of this species are marked with three wavy lines (Figure 101).

Life cycle

Eggs are laid amongst the foodstuff. Larvae hatch and initially feed on fragments and damaged grains, but as they get bigger they can attack whole grains. Larvae moult from 5 to 15 times (larger number when conditions are unfavourable). Cast skins remain entire and are a very obvious sign of the presence of these insects. Pupation occurs within the last larval skin. Adults are short lived and do not feed on the commodity. *T. granarium* cannot fly but other species fly well. Adult beetles do not feed on stored commodities but may visit flowers to feed on nectar and pollen.

Figure 93 *Trogoderma glabrum*, adult

Figure 94 *Trogoderma granarium*, adult, note relative lack of patterning on elytra

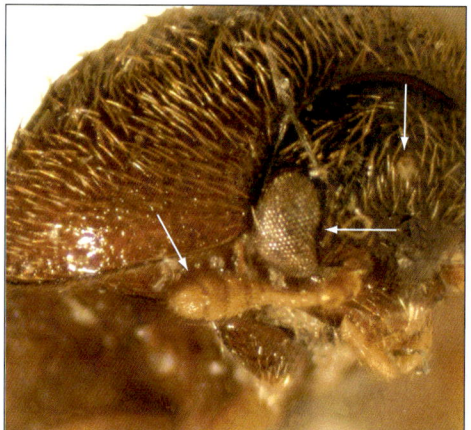

Figure 95 *Trogoderma granarium*, adult, head / antennae, showing median ocellus and antennal club with symmetrically jointed segments

Figure 96 *Trogoderma granarium*, adult, underside, showing antennal cavity

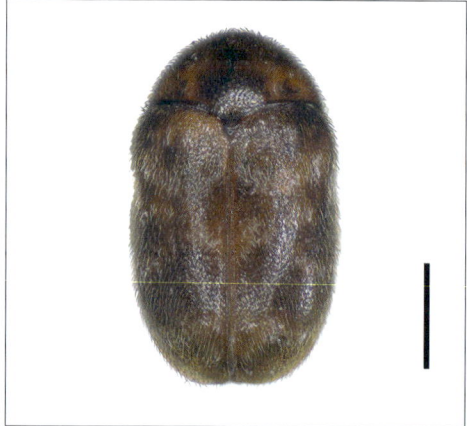

Figure 97 *Trogoderma inclusum*, adult

Figure 98 *Trogoderma inclusum*, adult, head, showing 'notch' in inner margin of eye

Figure 99 *Trogoderma* spp., adult, tarsus of hind leg, first and second segments of roughly equal length

Figure 100 *Trogoderma variabile*, adults, live, male (left) and female (right)

Figure 101 *Trogoderma variabile*, adult showing distinctive wavy bands on elytra

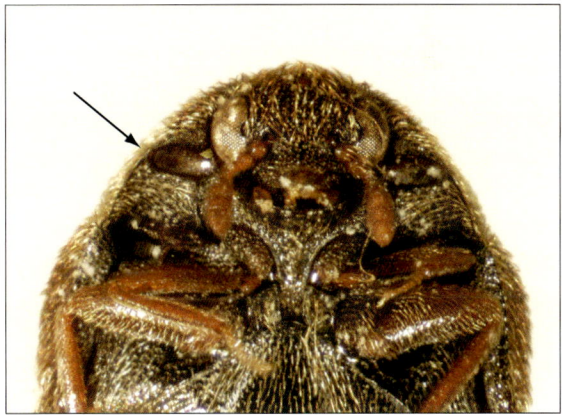

Figure 102 *Trogoderma variabile*, adult, head underside showing antennal cavity

Figure 103 *Trogoderma variabile*, larva, showing tufts of hairs including hastisetae on abdomen

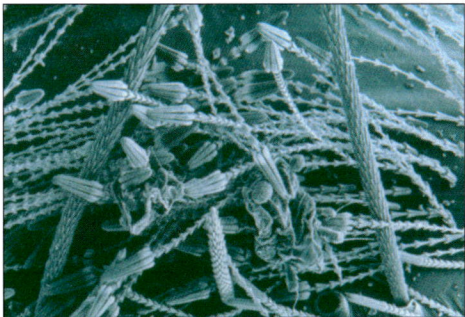

Figure 104 *Trogoderma variabile*, larva, hastisetae, showing their distinctive structure

Larvae of *Trogoderma* species are able to enter a state of type of suspended animation known as diapause. During this time, larvae hide away in cracks and do not feed, except for the occasional brief foraging excursion. In this state, larvae of *T. granarium* can live for up to eight years. This adaptation allows these insects to survive times when food is scarce or when environmental conditions are unfavourable. Diapause can be initiated by a number of things including lowered temperature, inadequate food and overcrowding.

Physical limits and optimum rate of multiplication

Species	Conditions within which breeding takes place	Shortest development period, with optimum conditions	Maximum monthly rate of increase
Trogoderma glabrum		30–36 days at 30°C, 65–70% r.h.	
Trogoderma granarium	20–40+°C, r.h. > 2%	25 days at 33–37°C, 45–75% r.h.	13
Trogoderma inclusum	20–40°C	60 days at 30°C.	
Trogoderma variabile	17–37°C	30 days at 30°C	8

T. granarium in particular is very tolerant of hot and dry conditions. Development time can greatly extended due to the ability of larvae to diapause.

Economic importance

By far the most important species worldwide are *T. granarium* and *T. variabile*. *T. granarium* is a feared pest of stored products and is especially serious on grain and oilseeds stored in bags in hot dry climates. Under such conditions, populations can cause total loss of infested commodities. Formerly it was a pest of maltings in Japan and northern Europe where it infested kilned grain or residues near the kilns. It also attacks a wide range of other plant-based materials, for example in retail packs of dried cucurbit seeds sold as snacks in the Middle East and south-west Asian region. Due to its ability to diapause, structural infestations of ships, rail cars and containers can persist for years after initial infestation, only to cross-infest other material at a later date. Used sacks can also be a source of infestation.

Next most important worldwide is *T. variabile,* which is often a difficult-to-eradicate pest in silos, mills and processing plants where it usually attacks crop residues. While not usually a serious pest of bulk grain, it can become a damaging pest of bagged grain and processed and packaged foods. In Australia, *T. variabile* has recently become a pest of bulk stored canola seed.

Larvae of *Trogoderma* spp. can penetrate most common packaging materials and commonly infest packaged foods of all kinds. They will infest rodent baits dispensed around store sites. They are also sometimes found attacking material of animal or plant origin in museums and private collections. As with other dermestids, bodies and cast skins of *Trogoderma* species are highly allergenic to humans.

Quarantine legislation exists in many countries to prevent the introduction of *Trogoderma* spp., especially *T. granarium.* Major exporters of agricultural commodities, such as the USA and Australia, take considerable effort to prevent the importation and establishment of this pest. A major reason for the widespread requirement to fumigate stored products in international trade is to prevent the spread of *T. granarium.*

Type of damage and symptoms

Trogoderma spp. are general feeders and as a result damage directly caused is not distinctive. However, if left unchecked, large quantities of cast larval skins can accumulate in and around infested materials.

Ecology

In laboratory studies, *Trogoderma* species breed most rapidly under conditions of moderate to high humidity; however, *T. granarium* is rarely found in such conditions. A reason for this appears to be competition with faster breeding pests such as *Sitophilus* and *Rhyzopertha*. Only when infesting hot dry grain does *T. granarium* have a competitive advantage.

The ability of larvae to diapause greatly enhances pest status, especially of *T. granarium*. In diapause, larvae can survive lengthy periods without food and are difficult to kill with pesticides and fumigants on account of their behaviour and low metabolic inactivity. When in this state they tend to hide in cracks and crevices making them difficult to find. They are able to survive extended periods in empty containers, ships' holds and other forms of transportation.

T. variabile is known as a pest of nests of the Alfalfa leaf cutting bee, an important pollinator in parts of North America. Natural habitats such as wasp nests and spider webs can serve as important reservoirs for populations of *Trogoderma* species generally.

In Australia, there are a large number of native species currently classified as *Trogoderma* species. Some of these occasionally find their way into grain storages. Little is known about their distribution, biology and potential pest status; many of them have yet to be named by the scientific community – see Banks (1994).

Monitoring

Infestations are usually spotted by accumulations of cast larval skins. A synthetic aggregation pheromone is commercially available. This pheromone is effective when used in flight traps to monitor *T. variable*. Care needs to be taken as the pheromone is not very specific and will attract other species of *Trogoderma* and members of closely related genera such as *Orphinus*. A crawling insect trap, containing a food bait and pheromone bait has been developed and used in the USA to monitor incursions of *T. granarium*.

Geographical distribution

Species	Pest status	USA & Canada	Central & South America	Europe & N.Asia	Mediterranean basin	Africa	S. & SE. Asia	Australia & Oceania
Trogoderma glabrum	●●	X	X®	X			X®	
Trogoderma granarium	●●●●			X®	X	X®	X®	
Trogoderma inclusum	●●	X	X	X	X			
Trogoderma ornatum	●●	X						
Trogoderma sternale	●●	X						
Trogoderma variabile	●●●	X	X®	X	X	X®	X	X

Pest status: ● minor to ●●●● major pest
X: recorded
®: restricted distribution

T. granarium probably originated from the north-west of the Indian sub-continent. It is now established in hot dry areas in a band from western Africa, through the northern half of Africa, southern and eastern Mediterranean, south-west Asia to northern India and Burma. Outbreaks have also been reported on the Malay peninsula (apparently eradicated), southern Philippines, Taiwan and Korea. It also appears to be spreading in the newly independent states of central Asia.

Quarantine officials in areas bordering these regions, such as in the southern and western provinces of China, are taking active measures to prevent the establishment and spread of this pest. In Japan and northern Europe *T. granarium* was formerly a pest of grain in heated buildings such as maltings. Due to changes in management practices and closure of old facilities, *T. granarium* has declined in these regions and is now either no longer established (e.g. Austria, UK) or has become very rare. *T. granarium* appears to be absent from the Americas, South Africa, Australia, New Zealand and Pacific islands. Formerly reported outbreaks in south-west USA and north-west Mexico (1950s), Venezuela (1980s) and South Africa (1950s) have either died out or have been eradicated.

T. variabile appears to have originated in central Asia. It is now found throughout the Northern Hemisphere, including southern Europe, south-west Asia, Russia, Mongolia, China, Canada and the USA. It is also established in southern Australia.

T. glabrum and *T. inclusum* are recorded from North America, Europe and northern and central Asia. *T. ornatum* and *T. sternale* are North American in distribution.

Histerid beetles
(Family: Histeridae)

Carcinops species
Saprinus species
Teretrius nigrescens

Summary

Feeding strategy	predator
Commodities attacked	other insects
Distribution	worldwide
Economic importance	beneficial insect
Eggs	laid amongst commodity
Larvae	active, campodeiform, with large forward-facing mandibles
Adults	long lived, feed, can fly

Introduction

Histerids are scavengers and predators in decomposing organic material such as animal carcasses. Other species are predators of wood-boring insects. A number of genera occur in the storage environment. *Carcinops* and *Saprinus* species are associated with the bodies of dead animals, including dried fish where they prey on fly larvae and other insects. Members of the genus *Teretrius* are predators on wood-boring beetles such as members of the Bostrichidae, Lyctinae and Scolytinae, hunting in the tunnels made by these pests. *Teretrius nigrescens* is a closely associated predator of the larger grain borer *Prostephanus truncatus* (Coleoptera: Bostrichidae).

Identification

Adult histerid beetles are oval, seed-like beetles, usually shiny black or dark metallic in colour (Figures 105–107). In all species, the elytra are short, leaving one or two abdominal segments exposed. The antennae have a spherical, three-segmented club. When alarmed, adult beetles tuck their legs close to their body and feign death.

Those that are found associated with stored products are 3–7 mm in length. Species such as *Carcinops,* that live in rotting material are typically flattened in cross-section. Species that live in tunnels bored by bostrichid and scolytid beetles, like *Teretrius* spp., are more cylindrical in cross-section. Larvae are elongate and campodeiform, those of *Teretrius* spp. grow up to 10 mm long. The head is heavily sclerotized, brown in colour with a pair of large, forward pointing sickle-shaped mandibles (Figure 106).

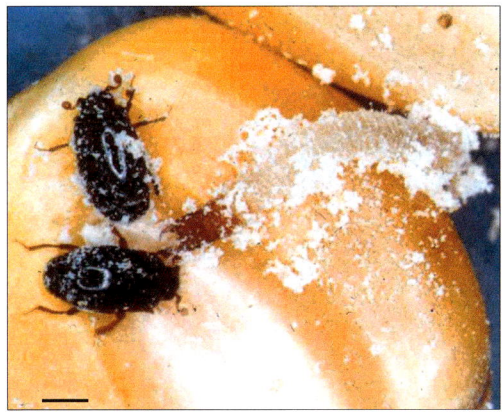

Top left: Figure 105 *Carcinops* species, adult, showing clubbed antennae and short elytra which leave tip of abdomen exposed

Top right: Figure 106 *Teretrius nigrescens*, adult and larvae, live

Left: Figure 107 *Teretrius nigrescens*, adult, showing short elytra which leave tip of abdomen exposed

Life cycle

Data is available for *Teretrius nigrescens*. Large eggs (about 1 mm long) are laid singly. At 27°C, 70% r.h., they hatch in about one week. Larvae hunt in the tunnels burrowed into the foodstuff by their prey. Larvae pass through two instars before pupation and prior to pupation they excavate a cell within the foodstuff or dried root. The pupal stage lasts for up to one month. Adults can live for over a year and can survive over six months without live prey, feeding on grain debris and flour. Reproduction is not possible without the presence of live prey.

Physical limits and optimum rate of multiplication

Species associated with rotting material are sensitive to low humidities. Species which attack wood-boring and related insects tolerate low humidities, such as those found in dry grain. Populations of *T. nigrescens* were observed to increase 3–3.5 times a month at 27°C, 70% r.h., when attacking populations of *P. truncatus*.

Economic importance

When present in large numbers, *Carcinops* and *Saprinus* species are indicators of poor storage conditions, for example the presence of dead rodents or birds or infestations of insects. *T. nigrescens* has been introduced into Africa from Mexico and Central America as a biological control agent to control *P. truncatus*.

Type of damage and symptoms

Both adult and larval histerids are for the most part predators. Adult *T. nigrescens* is known to feed on damaged grains in the absence of prey, but damage caused is not distinctive and is minor.

Ecology

T. nigrescens is very closely associated with the bostrichid *P. truncatus* both in the storage environment and in natural habitats. *T. nigrescens* is strongly attracted to aggregation pheromones produced by *P. truncatus*. and is thus able to rapidly locate populations of its prey, even if thinly and patchily spread.

Monitoring

Flight traps baited with synthetic *P. truncatus* aggregation pheromone will also attract *T. nigrescens*.

Geographical distribution

Histerid beetles are found throughout temperate and tropical regions of the world. Of species associated with stored products, *Carcinops* species are found worldwide. *T. nigrescens* is closely associated with *P. truncatus* in Mexico and Central America and is now established in parts of east and west Africa following its introduction. *Saprinus* species are associated with infestations of *Dermestes* on dried fish, mainly in tropical regions.

References

Aitken (1975), Arbogast (1991), Haines (1991) Hinton (1945a), Hodges (1986, 1994).

Flat grain beetles
(Family: Laemophloeidae – formally Family: Cucujidae)

Cryptolestes capensis	
Cryptolestes cornutus	
Cryptolestes divaricus	Flat grain beetle
Cryptolestes ferrugineus	Rust-red grain beetle
Cryptolestes klapperichi	
Cryptolestes pusillus	Flat grain beetle
Cryptolestes pusilloides	
Cryptolestes turcicus	
Cryptolestes ugandae	

Summary

Feeding strategy	secondary pest
Commodities attacked	grain and grain products, nuts, oilseeds, dried root crops
Distribution	worldwide
Economic importance	medium to high
Eggs	laid amongst commodity
Larvae	campodeiform, mobile, external feeders
Adults	long lived, feed on commodity, can fly

Introduction

The Laemophloeidae are small highly flattened beetles, the majority of which live under bark of trees and are of little economic importance. One genus, *Cryptolestes*, ranks among the most important pests of cereal and cereal products.

Identification

Cryptolestes spp. are small (1.5 to 2 mm long), reddish-brown, highly flattened parallel-sided beetles with long hair-like antennae (Figures 108–113). Head and prothorax together make up half the body length. When viewed from above, a ridge runs from above each eye down each side of the thorax (Figure 109). Antennae are long – up to the length of the body of the insect. Their highly flattened form, large head and thorax and long hair-like antennae distinguish *Cryptolestes* from other small beetles found in stored products. Unlike members of the similar Silvanidae (*Oryzaephilus, Ahasverus, Cathartus*), adult *Cryptolestes* cannot climb clean glass. Adult *Cryptolestes* species walk with a characteristic swaying motion.

Cryptolestes species are very similar to each other and identification by external characteristics alone is difficult. The most reliable method is to examine the genitalia of adult specimens. Specimens need to be cleared and slide mounted for microscopic examination, see Banks (1980) and Halstead (1993).

Figure 108 *Cryptolestes ferrugineus*, adult showing distinctive long antennae characteristic of genus

Figure 109 *Cryptolestes ferrugineus*, adult, head / thorax, showing ridge that runs from behind eye and across thorax

Figure 110 *Cryptolestes ferrugineus*, larva

Figure 111 *Cryptolestes ferrugineus*, infestation in rolled oats

Figure 112 *Cryptolestes pusillus*, adult

Figure 113 *Cryptolestes pusilloides*, adult

Life cycle

Eggs are laid in crevices in grains or loosely amongst the food. The flattened elongate campodeiform larva is active and moves freely through the commodity. Larvae, especially young ones, will enter minute cracks in the seed coat and burrow into the endosperm or germ. Pupation occurs either in a larval burrow in, or between grains. The long-lived adults feed on grain and grain products.

Physical limits and optimum rate of multiplication

Species	Conditions within which breeding takes place	Shortest development period, with optimum conditions	Maximum monthly rate of increase
Cryptolestes capensis	15–32.5°C, r.h. >10%	33 days at 30°C, 70% r.h.	
Cryptolestes ferrugineus	20–42.5°C, 40–90% r.h.	21 days at 35°C, 90% r.h	60
Cryptolestes pusillus	17.5–37.5°C, r.h. > 50%	22 days at 35°C, 90% r.h.	
Cryptolestes pusilloides	15–35°C, r.h. > 50%	27 days at 30°C, 90% r.h.	
Cryptolestes turcicus	17.5–35°C, r.h. > 50%	34 days at 35°C, 90% r.h.	50
Cryptolestes ugandae	17.5–35°C, r.h.> 50%	22 days at 25–27.5%, 90% r.h.	

Some strains of C. ferrugineus, C. capensis and C. turcicus are highly cold tolerant and able to survive extended periods at or below 0°C. This allows these species to be important pests in cool temperate areas. C. ferrugineus and C. capensis are able to breed under drier conditions than other species.

Economic importance

Cryptolestes are important pests of cereals, cereal products, oilseeds and dried processed foods of vegetable origin. Under tropical conditions they are found on a wider range of commodities including nuts, cocoa, copra and cassava. When infesting cereal-based products, Cryptolestes breed most rapidly on milled produce or on grain previously damaged by insects or by poor harvesting, storage or handling. They are often associated with heating grain. Being small and highly flattened, adults and larvae easily enter imperfections in packaged goods. Worldwide the most important species are C. ferrugineus and C. pusillus.

Type of damage and symptoms

Larvae and adults are general feeders, and damage is not readily identifiable as being specifically caused by this insect. Larvae feed preferentially on grain germ.

Ecology

Cryptolestes species often behave as secondary pests following infestations of primary stored grain pests such as Sitophilus spp. or Rhyzopertha dominica. However, C. ferrugineus can attack apparently sound grain with damage caused by harvesting and handling being sufficient to facilitate entry. This, combined with cold tolerance, means that in seasonally cold climates, such as the Canadian wheatbelt, this species is an important pest of stored cereal grain. In warmer regions of the world C. ferrugineus is frequently found with C. pusillus and sometimes

C. pusilloides. Under tropical conditions, *C. pusillus* can be the dominant species. In mills in some temperate regions, *C. turcicus* can replace *C. ferrugineus.* In Australia, *C. pusilloides* is frequently recorded as a pest of dried mushrooms.

Cryptolestes species can be found in a range of natural habitats such as under the bark of trees.

Monitoring

Cryptolestes species are easily caught in pitfall type traps inserted into commodity bulks. Crevice traps are also effective and their efficacy can be improved with addition of a food bait. A number of proprietary bait and trap systems are available which would be attractive to these pests.

Geographical distribution

Species	Pest status	USA & Canada	Central & South America	Europe & N.Asia	Mediterranean basin	Africa	S. & SE. Asia	Australia & Oceania
Cryptolestes capensis	●●			X	X			
Cryptolestes cornutus	●						X	
Cryptolestes divaricus	●						X	
Cryptolestes ferrugineus	●●●●	X	X	X	X	X	X	X
Cryptolestes klapperichi	●					X	X	
Cryptolestes pusillus	●●●	X	X	X	X	X	X	X
Cryptolestes pusilloides	●●	?	X	?	X	X	X	X
Cryptolestes turcicus	●●●	X	X	X	X		X	
Cryptolestes ugandae	●					X		

Pest status: ● minor to ●●●● major pest
X: recorded
?: status unclear

The genus *Cryptolestes* is cosmopolitan and a total of nine species are known from stored products. Of these only two – *C. ferrugineus* and *C. pusillus* are truly cosmopolitan. The rest currently have relatively restricted distributions and have potential for spread into new areas.

References
Aitken (1975), Arbogast (1991), Haines (1974, 1981, 1991) and Halstead (1993).

Minute mould beetles, Plaster beetles
(Families: Cryptophagidae and Latridiidae (Lathridiidae)) (selected genera)

Family: Cryptophagidae

Cryptophagus spp.

Family: Latridiidae

Aridius spp.

Cartodere spp.

Corticaria spp.

Latridius spp.

Microgramme spp.

Summary

Feeding strategy	mould feeder
Commodities attacked	damp material of vegetable origin including grain and grain products
Distribution	worldwide
Economic importance	low
Eggs	laid amongst commodity
Larvae	active, external feeders, campodeiform
Adults	long lived, feed, can fly

Introduction

Members of these families feed on fungi that grow on damp material of animal or plant origin. They can be found infesting damp or poorly stored commodities.

Identification

Compared to other commonly encountered species, these beetles are very small (1.3–2 mm long). Their size and general appearance alone distinguishes them as a group from other more important inhabitants of stored products.

Cryptophagus species are small hairy flat beetles. They have a characteristic tooth midway along the side edge or lateral margin of the pronotum (Figure 114). The head and thorax of members of the Latridiidae are small relative to the wider and more bulbous abdomen (Figures 115–118). The antennae have a two- or three-segmented club. Some species, e.g. *Corticaria* species, are hairy (Figures 116–117). Elytra of hairless species, e.g. *Latridius* spp., are marked with longitudinal ridges with rows of pits in between (Figure 118). Some species have a distinct lateral constriction about halfway along the pronotum (Figure 118).

Figure 114 *Cryptophagus* spp., adult, showing antennal club, teeth-like structures on margin of thorax and hairs on elytra

Figure 115 *Cartodere* spp., adult, showing antennal club, constriction in thorax and pitting on elytra

Above left: Figure 116 *Corticaria adelaidae*, adult, live

Above right: Figure 117 *Corticaria* spp., adult, showing antennal club, fine teeth-like structures on margins of thorax and hairs on elytra

Left: Figure 118 *Latridius* spp., adult, showing antennal club, constriction in thorax and pitting on elytra

Life cycle

Eggs are laid singly on the food. The flattened elongate campodeiform larva is active and moves freely through the commodity. Pupation occurs within the commodity.

Physical limits and optimum rate of multiplication

Data describing development of populations on stored products appears to be lacking. However, development of populations is reported for a number of species on cultures of moulds. Typically, they develop from egg to adult in 30–50 days at temperatures between 15–18°C.

Economic importance

Members of these families are mould feeders and they do not directly attack the stored commodity. Presence of large numbers of these insects indicates poor storage conditions or commodities which are damp and mouldy.

Type of damage and symptoms

Larvae and adults feed on mould growth and generally do not feed on stored commodities directly.

Ecology

These insects are common in ripening crops and may be found on grain at intake but usually do not breed or persist in dry commodities stored under good conditions.

Monitoring

Mould beetles are easily caught in pitfall type traps inserted into commodity bulks. Crevice traps are also effective.

Geographical distribution

Members of these families are found worldwide. However, they are most often noticed associated with stored products in temperate regions.

References

Aitken (1975), Bousquet (1990), Hinton (1941), Kingsolver and Andrews (1987).

Lophocaterid beetles
(Family: Lophocateridae – formerly included in the family Trogossitidae)

Lophocateres pusillus	Siamese grain beetle

Summary

Feeding strategy	scavenger
Commodities attacked	stored products of vegetable origin
Distribution	worldwide
Economic importance	low
Eggs	laid in amongst commodity
Larvae	campodeiform, mobile, external feeders
Adults	long lived, feed on commodity

Introduction

In nature lophocaterid beetles are found living under the bark of trees as scavengers. One species – *Lophocateres pusillus* – is associated with stored products.

Figure 119 *Lophocateres pusillus*, adult, showing highly flattened margins of thorax and elytra

Figure 120 *Lophocateres pusillus*, adult, underside, showing highly flattened margins of thorax and elytra

Identification

Adult *L. pusillus* are highly flattened parallel-sided beetles 3 mm long, and brown to dark grey in colour (Figure 119). The elytra are marked with longitudinal ridges. The sides of the pronotum are distinctively flattened to form flange-like structures (Figure 120).

Life cycle

Species	Conditions within which breeding takes place	Shortest development period, with optimum conditions	Maximum monthly rate of increase
Lophocateres pusillus	20–35°C at 75% r.h.	42 days at 30°C, > 10% r.h.	

Eggs of *L. pusillus* are laid amongst the commodity in cracks in the seed coat of grain. They are often laid in clumps of several or more. The flattened, campodeiform larvae live and feed amongst the commodity.

Economic importance

L. pusillus is a minor pest and scavenger in cereals, especially rice, beans and cassava, stored under tropical and sub-tropical conditions.

Type of damage and symptoms

Larvae and adults are general feeders, and damage is not readily identifiable as being specifically caused by this insect.

Ecology

This species often occurs in the presence of other storage pests, especially *Rhyzopertha*, *Sitophilus* and *Tribolium* species. Frass and dead insects appear to form an important part of its diet.

Monitoring

No specific information is available. However, this insect is likely to be captured by crevice or pitfall traps.

Geographical distribution

Species	Pest status	USA & Canada	Central & South America	Europe & N.Asia	Mediterranean basin	Africa	S. & SE. Asia	Australia & Oceania
Lophocateres pusillus	●		X			X	X	X

Pest status: ● minor to ●●●● major pest
X: recorded

L. pusillus is widely distributed in tropical regions and is a common inhabitant of rice stored in south and south-east Asia.

References
Aitken (1975), Arbogast (1991), Haines (1991).

Hairy fungus beetles
(Family: Mycetophagidae)

Litargus balteatus	
Mycetophagus quadriguttatus	Spotted hairy fungus beetle
Typhaea stercorea	Hairy fungus beetle

Summary

Feeding strategy	mould feeder
Commodities attacked	wide range of material of vegetable origin, including grain and grain products, especially if slightly damp
Distribution	worldwide
Economic importance	low
Eggs	laid in amongst commodity
Larvae	campodeiform, mobile, external feeders
Adults	long lived, feed on commodity, can fly

Introduction

The Mycetophagidae is a family of mould feeding beetles with some 200 described species. In nature, they are usually found associated with fungi growing on a wide range of plant material.

Identification

Mycetophagids are small flattened, somewhat hairy oval-shaped beetles 2–3 mm long, *M. quadriguttatus* and *L. balteatus* are distinctive on account of their patterned elytra (Figures

Figure 121 *Litargus balteatus*, adult, showing patterned elytra

Figure 122 *Typhaea stercorea*, adult, live

Figure 123 *Typhaea stercorea*, adult, showing antennal club and parallel lines of hairs on elytra

121–123). *T. stercorea* can be confused with other similar sized storage pests, in particular the tenebrionids *Palorus*, *Tribolium*, *Latheticus* and *Gnatocerus*. In comparison with *T. stercorea*, these beetles are not hairy and in shape are more parallel-sided.

Key to mycetophagids associated with stored products

1   Pronotum and elytra uniform brown in colour, flattened, oval, light brown, hairy beetle about 2–3 mm long. Antenna have three-segmented club. Hairs on elytra in parallel lines . . . . *Typhaea stercorea* (Figure 123)
Elytra bicoloured, with light patches on darker background 2

2   Antennal club with four segments, pair of deep oval pits near base of pronotum, length of adult 3 mm . . . . *Mycetophagus quadriguttatus*
Antennal club with three segments – last segment longer than other two segments, no deep oval pits near base of pronotum, length of adult 2.0–2.5 mm . . . .
. . . . *Litargus balteatus* (Figure 121)

Life cycle

Eggs of *T. stercorea* are laid at random or loosely attached to grains. The pale translucent larvae are very mobile and move through the commodity. Adults are long-lived, run very rapidly and can fly well. At 25°C, 80–90% r.h., development from egg to adult takes between 21 and 33 days.

Economic importance

These insects are minor pests of freshly harvested or slightly damp grain. In Australia for example *T. stercorea* is frequently encountered in baled hay presented for export.

Type of damage and symptoms

Larvae and adults are general feeders, and damage is not readily identifiable as being specifically caused by these insects.

Ecology

These insects can be found on ripening hay and grain crops before harvest and the presence of large and / or persistent populations in stores is likely to indicate poor storage conditions.

Monitoring

Mould beetles are easily caught in pitfall type traps inserted into commodity bulks. Crevice traps are also effective as may be food baits.

Geographical distribution

Species	Pest status	USA & Canada	Central & South America	Europe & N.Asia	Mediterranean basin	Africa	S. & SE. Asia	Australia & Oceania
Litargus balteatus	●	X	X	X	X	X	X	X
Mycetophagus quadriguttatus	●	X		X				
Typhaea stercorea	●	X	X	X	X	X	X	X

Pest status: ● minor to ●●●● major pest
X: recorded

L. balteatus and *T. stercorea* are found worldwide in both temperate and tropical areas. *M. quadriguttatus* appears to be restricted to temperate regions in the northern hemisphere. By far the most frequently encountered species is *T. stercorea*.

References

Aitken (1975), Arbogast (1991), Bousquet (1990), Haines (1991), Hinton (1945).

Dried fruit beetle, Corn sap beetles, Sap beetles, *Carpophilus* species

(Family: Nitidulidae) (selected species listed)

Carpophilus dimidiatus	Corn sap beetle
Carpophilus hemipterus	Dried fruit beetle
Carpophilus ligneus	
Carpophilus maculatus	
Carpophilus marginellus	
Carpophilus mutilatus	
Carpophilus obsoletus	
Carpophilus pilosellus	

Summary

Feeding strategies	secondary pest, mould feeder
Commodities attacked	dried fruit, damp and newly harvested grain, grain residues
Distribution	worldwide
Economic importance	variable – minor pest of grain, more important on dried fruit
Eggs	laid in crevices and folds of commodity
Larvae	campodeiform, active, external feeders
Adults	long lived, feed on commodity, fly readily

Introduction

The family Nitidulidae consists of more than 2000 described species. Many live on the sap of trees and juice of fruits, especially if partly fermented. Others feed on flowers, fungi and carrion. A few species are predatory and a few others are leaf miners.

A number of genera of nitidulids have been recorded as infesting stored products, including *Brachypeplus, Carpophilus, Glischrochilus, Haptoncus, Nitidula, Stelidota* and *Urophorus. Nitidula* are carrion feeders, and other genera listed are associated with fruit or other damp material of plant origin. By far the most important and frequently encountered genus on stored produce is *Carpophilus*.

Identification

Adult *Carpophilus* are oval, flattened beetles 2 to 4 mm long (Figures 124–127). Colour varies from light brown to black. Elytra are short and leave two or three abdominal segments exposed. Elytra are often marked with one or two yellow, reddish or brown spots. Antennae are terminated by a three-segmented flattened, round club. Most *Carpophilus* species are very similar in appearance and are difficult to identify to species. Many can only be reliably identified by microscopic examination of their genitalia and/or the texture of the cuticle of the underside of the insect (see References). However, one common species *C. hemipterus* (Figures 125–126) is easily distinguished on account of a characteristic large, roughly triangular, yellow spot on each elytra.

The general appearance and distinctive elytra of nitidulids distinguish *Carpophilus* spp. from other beetles commonly found in stored products. Members of the family Histeridae that occur in stores can sometimes be confused for dark specimens of *Carpophilus*. Histerids, which are predatory, also have clubbed antennae and short elytra which leave abdominal segments exposed. They are glossy black or metallic in colour and are not marked with yellow, reddish or brown spots.

Life cycle

Female beetles can lay 1000 eggs over a life span of 3–4 months. Eggs are laid on or in food material. The active translucent campodeiform larvae move amongst the commodity and may burrow into the soft and mouldy fruit or residue. Pupation occurs within the food material. The adults fly readily and often congregate at suitable feeding and oviposition sites.

Figure 124 *Carpophilus dimidiatus*, adult, showing short elytra which leave tip of abdomen exposed

Figure 125 *Carpophilus hemipterus*, adult, live

Figure 126 *Carpophilus hemipterus*, adult, showing short patterned elytra which leave tip of abdomen exposed

Figure 127 *Carpophilus marginellus*, adult

Physical limits to development

Species	Conditions within which breeding takes place	Shortest development period, with optimum conditions	Maximum monthly rate of increase
Carpophilus hemipterus	18.5–42°C, r.h.> 50%	12 days at 32°C, high humidity	50
Carpophilus dimidiatus	17.5–32.5°C, r.h. > 50%	15 days at 32°C, high humidity	

Warm damp conditions favour the rapid development of *Carpophilus* spp. Under optimal conditions and given a good food supply, development to adulthood can be extremely rapid.

Economic importance

Worldwide some several dozen species have been found associated with stored products. Among the most important are *C. hemipterus*, and *C. dimidiatus*. *Carpophilus* species will feed on a wide range of vegetable matter, especially if it is damp and decomposing. Usually only a minor pest of stored grain, *Carpophilus* spp. are common inhabitants of ripening cereal crops (they are often very common in ripening maize), and damp mouldy grain residues. *C. hemipterus* can be an important pest of dried fruit. A number of *Carpophilus* spp. are important pests of ripening soft and stone fruit.

Type of damage and symptoms
Larvae and adults are general feeders, and damage is not readily identifiable as being specifically caused by this insect. Larvae burrow into the soft and mouldy parts of grain, fruit or residues.

Ecology
Carpophilus spp. are attracted by moulds and yeasts on potential food and their presence in the diet appears very beneficial. Many species of *Carpophilus* including those associated with stored products are more important as pests of ripening fruit. They are highly mobile insects capable of rapid population development, well able to exploit transient habitats such as ripening grain or fruit. They are common inhabitants of compost heaps and other accumulations of rotting plant material.

Monitoring
In the orchard industry, flight traps of various designs have been used to trap *Carpophilus* species. These are baited with a synthetic aggregation pheromone and / or a food bait such as whole wheat bread dough or fermenting fruit juice.

Geographical distribution

Species	Pest status	USA & Canada	Central & South America	Europe & N.Asia	Mediterranean basin	Africa	S. & SE. Asia	Australia & Oceania
Carpophilus dimidiatus	••	X	X	X	X	X	X	X
Carpophilus hemipterus	•••	X	X	X	X	X	X	X
Carpophilus ligneus	•	X	X	X	X	X	X	X
Carpophilus maculatus	•		X			X	X	X
Carpophilus marginellus	•	X	X	X	X	X	X	X
Carpophilus mutilatus	•	X		X	X		X	X
Carpophilus obsoletus	•		X	X		X	X	X
Carpophilus pilosellus	•	X	X		X	X	X	X

Pest status: • minor to •••• major pest
X: recorded

Most species commonly found in stored products are cosmopolitan in distribution, being widespread in warm-temperate to tropical regions. In temperate regions, populations are most likely to persist in sheltered or heated environments.

References
Aitken (1975), Arbogast (1991), Connell (1987), Dobson (1952), Dobson (1954a), Dobson (1954b), Haines (1991), Hinton (1945b).

Silvanid beetles
(Family: Silvanidae)

Ahasverus advena	Foreign grain beetle
Cathartus quadricollis	Square-necked flour beetle
Monanus concinnulus	
Oryzaephilus acuminatus	
Oryzaephilus gibbosus	
Oryzaephilus mercator	Merchant grain beetle
Oryzaephilus surinamensis	Saw-toothed grain beetle

Summary

Feeding strategies	secondary pest, mould feeder
Commodities attacked	grain and grain products, oilseeds, nuts, herbs and spices, dried fruit
Distribution	worldwide
Economic importance	low to high
Eggs	laid in amongst commodity
Larvae	campodeiform, mobile, external feeders
Adults	long lived, feed on commodity, fly readily

Introduction

The Silvanidae are small flattened beetles, the majority of which live under bark of trees where they are mould and detritus feeders and sometime predators. *Ahasverus*, *Cathartus*, *Monanus* and *Oryzaephilus* species attack a wide range of stored products. *Oryzaephilus* species rank among the most important pests of stored products.

Identification

Silvanid beetles are small, highly flattened, parallel-sided beetles, 2.5 to 3.5 mm, long with tooth-like projections along the side and/or corners of the pronotum (Figures 128–135).

Figure 128 *Ahasverus advena*, adult, live

Figure 129 *Ahasverus advena*, adult, head, showing antennae and teeth-like structures at corners of thorax, margin of thorax curved

Figure 130 *Cathartus quadricollis*, adult, showing square corners of thorax

Figure 131 *Monanus* spp., adult, showing patterned elytra (non-storage species shown)

Figure 132 *Oryzaephilus mercator*, adult, head and thorax, showing teeth-like structures and three longitudinal ridges on thorax, length of area behind eye relatively short

Figure 133 *Oryzaephilus surinamensis*, adult

Figure 134 *Oryzaephilus surinamensis*, adult, head, showing teeth-like structures and three longitudinal ridges on thorax, length of area behind eye relatively long

Figure 135 *Oryzaephilus surinamensis*, larva

Key to the major silvanids associated with stored products

1 Side of prothorax decorated by six tooth-like projections (Figure 132), colour dark brown to dark grey (*Oryzaephilus* species) (Figures 132–134) 2
Sides of prothorax without six tooth-like projections, colour light brown (Figures 128–131) 3

2 Length of the temple – the area of the head behind the eye long (Figure 134) . . . .
. . . . *Oryzaephilus surinamensis*
Length of the temple – the area of the head behind the eye short (Figure 132) . . . .
. . . . *Oryzaephilus mercator*

Elytra marked midway across with a broad brown to black lateral patch or band (Figure 131) . . . . *Monanus* spp.

Elytra not so marked 4

3 Pronotum with somewhat curved sides and an obvious tooth like structure at each front angle, sides of abdomen somewhat curved (Figure 129) . . . . *Ahasverus advena*
Pronotum with parallel sides, angles of pronotum square, sides of abdomen straight and parallel-sided (Figure 130) . . . . *Cathartus quadricollis*

Life cycle

The eggs are laid loose amongst the substrate or in cracks and crevices in grains. Several hundred are laid over the life span of the female. The flattened white to pale yellow campodeiform larvae (Figure 135) move freely amongst the foodstuff and eventually pupate within a cocoon-like structure made from small grains or food particles. Adult silvanids are long-lived and continue to feed during their lives. For example at 30°C, adult *O. surinamensis* live on average 6–8 months, but under cool temperate conditions this may extend to several years.

Physical limits and optimum rate of multiplication

Species	Conditions within which breeding takes place	Shortest development period, with optimum conditions	Maximum monthly rate of increase
Ahasverus advena	> 17.5°C, > 65% r.h.	22.5 days at 27°C, 75% r.h.	
Cathartus quadricollis	20–30°C, r.h. > 65%	20 days at 27–28.5°C, 80–85% r.h.	
Oryzaephilus surinamensis	20–38°C, r.h. > 10%	20 days at 30–32.5°C, 70–90% r.h.	50
Oryzaephilus mercator	18–38°C, r.h. > 10%	25 days at 30–32.5°C, 70% r.h.	20
Oryzaephilus acuminatus	20–37.5°C, r.h. > 30%	18 days at 32.5°C, 80–90% r.h.	

Populations of *O. surinamensis* from temperate areas are often highly cold tolerant and are able to survive extended periods at or below 0°C. Both *O. surinamensis* and *O. mercator* are capable of breeding under dry conditions. In contrast, *A. advena* and *C. quadricollis* are sensitive to low humidities.

Economic importance

A. advena is a minor secondary pest of a wide range of produce including cereal grains and products, oilseeds, copra, groundnuts, dried fruit, dried herbs and cocoa beans. It is often present at harvest but usually does not persist in dry clean grain. Large, persistent populations usually indicate unsatisfactory storage conditions with the presence of mouldy grain. In the USA, *A. advena* is also a nuisance pest associated with damp and newly constructed houses.

C. quadricollis is best known as a pest of maize. Infestation often begins in the field in ripening cobs and continues in storage. Infestations can be severe, especially in maize stored on the cob under conditions of tropical subsistence agriculture. *M. concinnulus* is a minor pest on a wide range of stored products.

O. surinamensis is an important pest of stored cereals, particularly milled and processed products. It also occurs on a very wide range of other commodities including dried fruit, nuts and oilseeds. In temperate regions, *O. surinamemsis* is often regarded as the most important insect pest of stored cereals. Its close relative *O. mercator* is more often found on commodities such as dried fruit and oilseeds rather than cereals. It is also an important pest, but generally less so than *O. surinamensis*. The small size and flattened form allow *Oryzaephilus* spp. to easily enter packaged goods. *O. gibbosus* has been found infesting coconut shell, oil palm fruits and groundnuts. *O. acuminatus* has been found infesting coconut shell and dried neem seed.

Type of damage and symptoms

Larvae and adults are general feeders, and damage caused is not readily identifiable as being specifically caused by these insects.

Ecology

A. advena and *C. quadricollis* can frequently be found in ripening crops. Both persist better in stored crops if they are stored in poor condition and when some mould growth is present.

In cereal grain, *O. surinamensis* is a secondary pest which can follow infestations of primary pests such as *Sitophilus* spp. or *Rhyzopertha dominica*. It has difficulty in penetrating whole and unbroken grains, however, it can easily enter kernels that have only slight imperfections, which are invariably present in commercially handled grain. The larvae and adults will feed preferentially on the germ but larvae can develop on endosperm alone. Milled produce or grain damaged by other insects is very susceptible to attack. *O. surinamensis* is known to supplement its diet by feeding on eggs and dead bodies of stored product moths.

Silvanid beetles can fly well, however, *Oryzaephilus* species do not fly as readily as *A. advena* or *C. quadricollis,* but will readily walk long distances. These beetles are easily able to climb up smooth surfaces such as glass.

Monitoring

Silvanid beetles are active mobile insects that are easily captured in pitfall type traps inserted into grain. Collection vessels have to contain either an oil or similar to trap insects or be coated with fluon (PTFE emulsion) to prevent insects from climbing out. Silvanids also respond well to food baits such as chopped carob; in addition, several commercial baits based on grain oils are available.

Geographical distribution

Species	Pest status	USA & Canada	Central & South America	Europe & N.Asia	Mediterranean basin	Africa	S. & SE. Asia	Australia & Oceania
Ahasverus advena	●●	X	X	X	X	X	X	X
Cathartus quadricollis	●●	X®	X	X	X	X	X®	
Monanus concinnulus	●	?	X			X	X	
Oryzaephilus acuminatus	●						X	
Oryzaephilus gibbosus	●					X		
Oryzaephilus mercator	●●●	X	X	X	X	X	X	X
Oryzaephilus surinamensis	●●●●	X	X	X	X	X	X	X

Pest status: ● minor to ●●●● major pest
X: recorded
®: restricted distribution
?: status unclear

A. advena, O. surinamensis and *O. mercator* are found worldwide. *A. advena* and *O. surinamensis* are cold-hardy and are common in unheated locations in cool temperate regions such as northern Europe. *O. acuminatus* and *O. gibbosus* are little known tropical species. *C. quadricollis* is known from maize growing areas in southern USA, Mexico, Central America and tropical South America, where it is probably native. It has also been introduced into maize growing regions of west Africa. It has also been recorded from Thailand and Borneo. *M. concinnulus* is widely distributed in tropical regions and is intercepted in temperate regions from time to time on imported products.

References
Aitken (1975), Arbogast (1991), Haines (1991), Halstead (1993).

Tenebrionid beetles, Darkling beetles
(Family: Tenebrionidae)

Introduction
The Tenebrionidae is a large family of beetles with about 10 000 described species. In nature, they are scavengers of plant and animal debris and some are partly predacious. They are often found under bark of trees and in tunnels bored by wood-boring beetles. About 100 species have been reported from the storage environment. Most are associated with residues and crops held in poor, especially damp conditions. A few, however, are adapted to cope with dry conditions and some, notably *Tribolium* spp., are amongst the most important pests of stored products.

Adults of most species associated with stored products are between 3–10 mm in length. *Tenebrio* and *Blaps* are bigger. Most are similar in appearance, being hairless, flattened, parallel-sided beetles reddish brown to black in colour. *Alphitobius* are more oval in shape and *Blaps* are exceptional being almost globular. The elytra of these beetles always completely cover the abdomen. Antennae are fairly short and either simple or clubbed at the tip. In most species when viewed from the side the eye of the adult is divided with a backward projection of the head. Larvae are more or less similar being elongate, largely hairless, often with light and dark bands and leathery in appearance (elateriform).

Major genera of Tenebrionidae associated with stored products can be identified by the following key. By far the most frequently encountered genus in stored products are *Tribolium*.

Key to major genera of Tenebrionidae associated with stored products
1   Large beetles – more than 12 mm in length 2   Smaller beetles – less than 8 mm 3
2   Flattened, parallel-sided, 12–15 mm long . . . . *Tenebrio* (Figures 164–167)   Globular, with long spidery legs 20–35 mm long . . . . *Blaps* (Figure 141)

3   Oval shaped or broad, sides of elytra somewhat curved, 4.5–7 mm in length 4
      Parallel-sided and narrow, sides of elytra straight, most species less than 4.5 mm in
      length 7

4   When viewed from below pseudopleuron ('flange' at outer edge of elytra) abruptly
    narrowed before tip of abdomen (Figure 163) . . . . *Sitophagus* (Figures 162, 163)
    When viewed from below pseudopleuron gradually narrows towards tip of abdomen
    (Figure 145) 5

5   Head widest at eyes (Figure 144) 6
    Head widest before eyes (Figure 137) . . . . *Alphitobius* (Figures 136–140)

6   Eyes large, from above distance between them about 33% of head width . . . . *Palembus*
    Eyes small, from above distance between them greater than 50% of head width (Figure
    144) , ridge of pseudopluron easily visible from above (Figure 145) . . . .
    . . . . *Cyaenus* (Figures 143–145)

7   Eyes entire and round when viewed from side (Figure 158) . . . .
    . . . .*Palorus* (Figures 155–160), *Coeleopalorus* (Figure 161)
    Eye divided across middle by side margins of head 8

8   Antennae with final segment much narrower than others before it (Figure 154) . . . . . .
    . . . . *Latheticus* (Figures 153, 154)
    Antennae with final segment same width or wider than segments before it (Figure 173)
    9

9   Prosternal process (between bases of front legs) parallel-sided and pointed at tip, male
    specimens with obviously enlarged mandibles (Figures 148, 149, 152) . . . .
    . . . . *Gnatocerus* (Figures 146–152)
    Prosternal process (between bases of front legs) board and widest at tip (shaped like axe
    head), males without obvious horns on mandibles (Figure 173, 177) . . . .
    . . . . *Tribolium* (Figures 168–184)

References

Aitken (1975), Arbogast (1991), Haines (1991), Mound (1989), Spilman (1987b), Sokoloff (1972).

Lesser mealworms, Black fungus beetle
(Genus: *Alphitobius*)

Alphitobius diaperinus	Lesser mealworm
Alphitobius laevigatus	Lesser mealworm, Black fungus beetle

Summary

Feeding strategies	secondary pest, mould feeder, scavenger
Commodities attacked	damp and mouldy plant material, animal and plant origin
Distribution	worldwide
Economic importance	low–medium
Eggs	laid amongst commodity
Larvae	elateriform, mobile, live amongst commodity
Adults	long lived, feed on commodity, can fly

Introduction

Members of the genus *Alphitobius* are scavengers, part-time predators and fungus feeders and in nature are encountered in decaying organic material and under the bark of trees. Two species, *A. diaperinus* and *A. laevigatus*, are frequently encountered as pests of stored produce especially if damp and are also found in animal houses feeding on faeces and dead animals.

Identification

Adult *Alphitobius* are 5.5 to 7 mm long, oval, and flattened (Figures 136–140). They are reddish brown to black depending on age and species. Larvae are elateriform, brown and about 15 mm long when fully grown. Adults of the two species can be distinguished from each other as below.

Identification of *Alphitobius* species associated with stored products
Eye when viewed from side is divided, minimum number of eye facets at narrowest point is 3 to 4 (Figure 138) . . . . *Alphitobius diaperinus*
Eye when viewed from side is almost completely divided, minimum number of eye facets at narrowest point is 1 (Figure 140) . . . . *Alphitobius laevigatus*

Life cycle

Eggs are laid in clumps and are stuck onto the foodstuff or other substrate. Eggs hatch in three to 10 days. Larvae are active and range through the foodstuff. When ready to pupate, they fashion a chamber in the food medium. Pupation takes place within this chamber. Adults are long-lived, feed and can fly. Mean adult life span at 21–24°C can be over 400 days. Upon hatching, adults go through a pre-oviposition period of about 13 days, after which females lay about four eggs a day for much of their remaining life.

Top left: Figure 136 *Alphitobius diaperinus*, adult, live

Top right: Figure 137 *Alphitobius diaperinus*, adult, head from above, showing widest point of head before eyes

Left: Figure 138 *Alphitobius diaperinus*, adult, head and eye from side, showing narrowest part of eye three or four facets wide

Figure 139 *Alphitobius laevigatus*, adult

Figure 140 *Alphitobius laevigatus*, adult, head and eye from side, showing narrowest part of eye one facet wide

Physical limits and optimum rate of multiplication

Species	Conditions within which breeding takes place	Shortest development period, with optimum conditions	Maximum monthly rate of increase
Alphitobius diaperinus		46 days at 32°C, 95% r.h.	

A. diaperinus is cold-tolerant and can survive ambient winter conditions in cool temperate climates.

Economic importance

A. diaperinus and *A. laevigatus* are scavengers, mould feeders and minor pests of a wide range of cereals and cereal products, especially if damp. They are found in residues and sweepings and not usually in clean dry grain. *A. diaperinus* is a common inhabitant of intensive poultry houses where it helps to compost the large amount of manure produced by the birds. It will also feed on the dried carcasses of animals (mice, birds etc.).

Type of damage and symptoms

Larvae and adult *Alphitobius* spp. are general feeders and damage is not readily identifiable as being specifically caused by this insect. Infestation can lead to persistent disagreeable odours in the commodity due to secretion of benzoquinones from abdominal glands.

Ecology

In grain storage large populations usually only occur under conditions of poor storage and hygiene. Larvae develop best under conditions of high humidity and when mould is present in the diet.

Monitoring

Alphitobius spp. are easily caught in pitfall type traps. Crevice traps are also effective and their efficacy can be improved with addition of food bait. A number of proprietary bait and trap systems are available which would be attractive to these pests.

Geographical distribution

Species	Pest status	USA & Canada	Central & South America	Europe & N.Asia	Mediterranean basin	Africa	S. & SE. Asia	Australia & Oceania
Alphitobius diaperinus	●●	X	X	X	X	X	X	X
Alphitobius laevigatus	●●	X	X	X	X	X	X	X

Pest status: ● minor to ●●●● major pest
X: recorded

Both species are cosmopolitan. *A. diaperinus* is perhaps more frequently encountered in temperate regions.

Churchyard beetles, Egyptian beetle (*Blaps* spp.)

Summary

Feeding strategy	scavenger
Commodities attacked	grain residues, animal faeces
Distribution	Europe, Mediterranean region, south-west Asia, introduced elsewhere, e.g. Australia
Economic importance	low
Eggs	laid amongst commodity
Larvae	elateriform, mobile, live amongst commodity
Adults	long lived, feed on commodity, cannot fly

Introduction
The genus *Blaps* occur in North Africa, Europe and Asia, with a distribution centred on the Mediterranean basin and south-west Asia. They are carrion feeders and scavengers of materials such as dried animal faeces. Several species have been accidentally introduced elsewhere.

Identification
Adult beetles are very distinctive (Figure 141–142). They are large (20–35 mm long), black, globular beetles with long spider-like legs.

10 mm

Figure 141 *Blaps* spp., adult, live

Figure 142 *Blaps* spp., mating pair

Life cycle

Eggs are laid at random as females move through the food media. Larvae are active and move through the food and pupation occurs within a cell made in the foodstuff. Adults are long lived, feed and cannot fly.

Physical limits and optimum rate of multiplication

No specific data are available – population development is likely to be slow. Some species are native to cool temperate regions and are thus cold-tolerant. Adults are very long-lived.

Economic importance

Blaps spp. are scavengers of minor importance. Their presence may indicate the presence of unhygienic conditions. For example in parts of southern Australia, the introduced *Blaps polycresta* is a nuisance pest in and around homesteads and grain storage facilities, where it feeds on residues and animal faeces.

Type of damage and symptoms

When disturbed, these beetles emit toxic and corrosive secretions from glands in their abdomen. Injuries have occurred as a result of people handling them.

Ecology

In nature, *Blaps* species are scavengers feeding on decaying plant material and faeces of animals.

Monitoring

Blaps spp. are easily caught in pitfall type traps. They are large obvious insects unlikely to be missed.

Geographical distribution

Species	Pest status	USA & Canada	Central & South America	Europe & N.Asia	Mediterranean basin	Africa	S. & SE. Asia	Australia & Oceania
Blaps spp.	●	X		X	X			X

Pest status: ● minor to ●●●● major pest
X: recorded

Blaps spp. occur naturally in Europe, Mediterranean basin and south-west Asia, especially in arid and semi-arid areas. *Blaps* have been accidentally imported into North America and southern Australia.

Larger black flour beetle (*Cynaeus angustus*)

Summary

Feeding strategies	primary pest, secondary pest, mould feeder
Commodities attacked	grain and grain products, especially maize
Distribution	North America, Europe?
Economic importance	medium
Eggs	laid amongst commodity
Larvae	elateriform, mobile, live amongst commodity
Adults	long lived, feed on commodity, can fly

Introduction

C. angustus is found under bark and around the base of Yucca plants as a scavenger and part-time predator and was restricted to the south-west USA and north-west Mexico. Recently, it has also become a pest of stored grain and as a result is spreading in its distribution.

Identifiction

Adult beetles are 5–6 mm long, dark brown to black and slightly oval in shape (Figure 143–145). The elytra relative to other storage tenebrionids are strongly ridged (Figure 145). The outside edge of the elytra (pseudopleuron) is flattened and easily seen from above (Figure 145). Larvae are elateriform, 12–15 mm long when fully grown and light brown in colour.

Figure 143 *Cyaneus angustus*, adult

Figure 144 *Cynaeus angustus*, adult head from above, showing head widest at eyes

Figure 145 *Cynaeus angustus*, adult, margin of elytra (pseudopleuron) obviously flattened and gradually narrows towards tip.

Life cycle

Eggs are laid at random as females move through the food media. Females may lay up to 350–450 eggs over their lifetime of a year or more. Larvae are active and move through the food and pupation occurs within a cell made in the foodstuff. High moisture content and presence of some mould is conducive to rapid development of this pest. Adults are long-lived, feed and can fly.

Physical limits and optimum rate of multiplication

Species	Conditions within which breeding takes place	Shortest development period, with optimum conditions	Maximum monthly rate of increase
Cyaneus angustus		35–40 days at 30°C	

Economic importance

C. angustus is a pest of increasing importance, especially of stored shelled maize. While this pest breeds most rapidly on damaged grain it is capable of attacking whole grain. It has also been found in small numbers as an inhabitant of chicken houses.

Type of damage and symptoms

Larvae and adult C. angustus are general feeders and damage is not readily identifiable as being specifically caused by this insect. Infestation can lead to persistent disagreeable odours in the commodity due to secretion of benzoquinones from abdominal glands.

Ecology

Prior to the 1930s, C. angustus was only known from natural habitats in south-west USA and neighbouring parts of Mexico. In nature, it is an inhabitant of the flour stem and debris at base of yucca plants. While it is not known exactly when this species became a pest of stored produce it is clear that it has only recently become one.

Monitoring

C. angustus are easily caught in pitfall type traps. Crevice traps are also effective and their efficacy can be improved with addition of food bait. A number of proprietary bait and trap systems are available which would be attractive to these pests.

Geographical distribution

Species	Pest status	USA & Canada	Central & South America	Europe & N.Asia	Mediterranean basin	Africa	S. & SE. Asia	Australia & Oceania
Cynaeus angustus	●●	X	X®	X®?				

Pest status: ● minor to ●●●● major pest
X: recorded
®: restricted distribution
?: status unclear

This insect still has a restricted distribution, but it is spreading. In North America, it has reached as far north as southern Canada and in the 1990s it had been found for the first time in Sweden. It has potential to become established in other temperate areas and is a species to watch.

Horned flour beetles (Genus: *Gnatocerus*)

Gnatocerus cornutus	Broadhorned flour beetle
Gnatocerus maxillosis	Slenderhorned flour beetle

Summary

Feeding strategy	secondary pest
Commodities attacked	dried material of animal and plant origin
Distribution	worldwide, *G. maxillosis* mainly tropical
Economic importance	low–medium
Eggs	laid amongst commodity
Larvae	elateriform, mobile, live amongst commodity
Adults	long lived, feed on commodity, can fly

Introduction

Members of the genus *Gnatocerus* are found under the bark of tree as scavengers and part-time predators. Two species, *G. cornutus* and *G. maxillosis*, are minor pests of stored products of both animal and plant origin.

Identification

Gnatocerus spp. are flattened, parallel-sided, reddish-brown beetles 3–4 mm long (Figures 146–152). Mandibles of males are obviously enlarged (see below). It is not easy to identify female specimens of *Gnatocerus* to species, except by association with males. Larvae are active, elateriform and light brown in colour (Figure 150).

Figure 146 *Gnatocerus cornutus*, adult, male, live

Figure 147 *Gnatocerus cornutus*, adult, female, live

Figure 148 *Gnatocerus cornutus*, adult, head of male showing enlarged side of head and mandibles, process between front legs parallel-sided

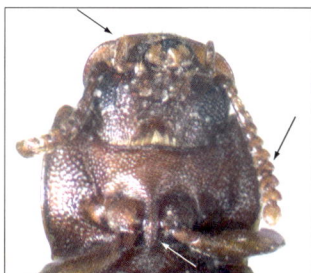

Figure 149 *Gnatocerus cornutus*, adult, head of female, side of head and mandibles not expanded, antennae without obvious club, process between front legs parallel-sided

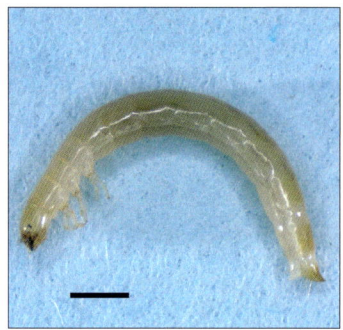

Figure 150 *Gnatocerus cornutus*, larva

Figure 151 *Gnatocerus maxillosis*, adult

Figure 152 *Gnatocerus maxillosis*, adult, head showing expanded mandibles of male

Identification of *Gnatocerus* species associated with stored products

Mandibular horns of male large and obvious, sides of head enlarged (Figure 148) . . . . *Gnatocerus cornutus*

Mandibular horn of male small, sides of head not enlarged (Figure 152) . . . . *Gnatocerus maxillosis*

Life cycle

Eggs are laid freely amongst the commodity, and up to 1200 can be laid by a female over a 7–14 month life span. Larvae live freely amongst food. Pupae are found amongst the food, as no pupal cell is made. The long-lived adults feed and can fly. The presence of animal protein in the diet improves fecundity. They are partly predatory, feeding on both grain material and other insects both living and dead.

Physical limits and optimum rate of multiplication

Species	Conditions within which breeding takes place	Shortest development period, with optimum conditions	Maximum monthly rate of increase
Gnatocerus cornutus	16–32°C, r.h. > 5%	57 days at 24–30°C, 66–92% r.h.	15
Gnatocerus maxillosis	17.5–35°C, r.h. > 7%	30–40 days at 30°C	

Both species are tolerant of a wide range of climatic conditions and low humidity. Population growth is slow relative to *Tribolium* species.

Economic importance

Gnatocerus spp. are minor pests of flour and feed mills and are scavengers on grain debris. Presence of large numbers of these insects suggests poor hygiene and presence of aged residues. Both species can become relatively important pests of grain when stored under conditions of tropical subsistence agriculture.

Type of damage and symptoms

Larvae and adult *Gnatocerus* spp. are general feeders and damage is not readily identifiable as being specifically caused by this insect. Infestation can lead to persistent disagreeable odours in the commodity due to secretion of benzoquinones from abdominal glands.

Ecology

Owing to its relatively slow rate of reproduction, *Gnatocerus* may be out-competed by more rapidly reproducing species such as *Tribolium*. *Gnatocerus* populations thrive best when they can prey on other insects. They also thrive on commodities which have a high proportion of protein, such as fish meal and dried material of animal origin.

Monitoring

Gnatocerus spp. are easily caught in pitfall type traps. Crevice traps are also effective and their efficacy can be improved with addition of food bait. A number of proprietary bait and trap systems are available which would be attractive to these pests.

Geographical distribution

Species	Pest status	USA & Canada	Central & South America	Europe & N.Asia	Mediterranean basin	Africa	S. & SE. Asia	Australia & Oceania
Gnatocerus cornutus	●●	X	X	X	X	X	X	X
Gnatocerus maxillosis	●●	X	X	X	X	X	X	X

Pest status: ● minor to ●●●● major pest
X: recorded

Both species are found worldwide. *G. cornutus* is more likely to be found in temperate regions and *G. maxillosis* in tropical areas. Both are not cold-hardy and will only survive winter in cool temperate climates in heated premises.

Longheaded flour beetle (*Latheticus oryzae*)

Summary

Feeding strategy	secondary pest
Commodities attacked	any dried material of animal and plant origin
Distribution	worldwide
Economic importance	low–medium
Eggs	laid amongst commodity
Larvae	elateriform, mobile, live amongst commodity
Adults	long lived, feed on commodity, can fly

Introduction

Members of the genus *Latheticus* lives in rotten wood and under bark as scavengers and part-time predators. One species, *L. oryzae*, is a minor pest of stored grain.

Identification

L. oryzae is a slender, flattened, parallel-sided, light yellowish to brown beetle of about 3 mm long (Figures 153–154). When viewed from the side, eyes are crescent shaped. Its antennae are distinctive, short, with a five-segmented club, the final segment of which is narrower than the one preceding it (Figure 154). Larvae are active, elateriform, and light brown.

Figure 153 *Latheticus oryzae*, adult, live

Figure 154 *Latheticus oryzae*, adult, head showing last segment of antennae narrower than segments before it

Life cycle

Eggs are laid at random as females move through the food media. They are sticky and become coated with flour or other particles. Adults are long-lived, feed and can fly. Females may lay up to 300 eggs during their lifetime of three to four months. Larvae are active and move through the food like the adults. Pupae are naked and are also found amongst the food.

Physical limits and optimum rate of multiplication

Species	Conditions within which breeding takes place	Shortest development period, with optimum conditions	Maximum monthly rate of increase
Latheticus oryzae	25–40°C, r.h. > 30%	22 days at 35°C, 85% r.h.	10

Compared to most other storage insects, *L. oryzae* has an unusually high minimum temperature required for development to take place.

Economic importance

Normally, *L. oryzae* is a minor pest of cereals and cereal products and oilseeds, especially under the hot conditions of the humid tropics or elsewhere in heated grain. It is a common inhabitant of flour mills in hot climates. It is usually a much less important pest than *Tribolium*, but can become dominant in hot and heating grain.

Type of damage and symptoms

Larvae and adult *L. oryzae* are general feeders and damage is not readily identifiable as being specifically caused by this insect. Infestation can lead to persistent disagreeable odours in the commodity due to secretion of benzoquinones from abdominal glands.

Ecology

Under normal conditions of storage, *L. oryzae* is likely to be out-competed by faster breeding species such as *Tribolium castaneum*. However, due to its tolerance to high temperatures, *L. oryzae* may become dominant in heating grain. *L. oryzae* is unable to attack sound whole cereal grains, but even very small imperfections allow access by young larvae. Presence of damaged grains, grain germ, moulds and frass of other insect species is beneficial to growth of populations of this insect.

Monitoring

L. oryzae are easily caught in pitfall type traps. Crevice traps are also effective and their efficacy can be improved with addition of food bait. A number of proprietary bait and trap systems are available which would be attractive to these pests.

Geographical distribution

Species	Pest status	USA & Canada	Central & South America	Europe & N.Asia	Mediterranean basin	Africa	S. & SE. Asia	Australia & Oceania
Latheticus oryzae	●●	X	X	X	X	X	X	X

Pest status: ● minor to ●●●● major pest
X: recorded

Found worldwide but most abundant in the tropics, in particular Asia. It is susceptible to cold conditions and in temperate regions is usually restricted to heated premises or to hot grain, e.g. that heated by drying or by the metabolic activities of other pests.

Palembus (Ulomoides) spp.

Summary

Feeding strategies	secondary pest, mould feeder
Commodities attacked	dried material of vegetable origin
Distribution	tropical, especially Asia
Economic importance	low
Eggs	laid amongst commodity
Larvae	elateriform, mobile, live amongst commodity
Adults	long lived, feed on commodity, can fly

Introduction

One species of this genus, *P. dermestoides*, is of note as it is reared and often consumed alive as a traditional Chinese medicine.

Identification

Adult beetles are 4.5–7 mm in length, oval in shape with the sides of the elytra somewhat curved and dark brown to black in colour.

Life cycle

Eggs are laid at random as females move through the food media. Females may lay up to 800 eggs over an oviposition period of about 100 days. Larvae are active and move through the food. Adults are long lived.

Physical limits and optimum rate of multiplication

Species	Conditions within which breeding takes place	Shortest development period, with optimum conditions	Maximum monthly rate of increase
Palembus dermestoides	> 18°C	42 days at 30°C	

Economic importance

Palembus spp. are minor pests of a wide range of stored produce. One species, *P. dermestoides*, is of interest as it is reared as a traditional medicine by some Asian communities. Beetles are apparently eaten alive.

Type of damage and symptoms

Larvae and adult *Palembus* spp. are general feeders and damage is not readily identifiable as being specifically caused by this insect. Infestation can lead to persistent disagreeable odours in the commodity due to secretion of benzoquinones from abdominal glands.

Monitoring

Palembus spp. are easily caught in pitfall type traps. Crevice traps are also effective and their efficacy can be improved with addition of food bait.

Geographical distribution

Species	Pest status	USA & Canada	Central & South America	Europe & N.Asia	Mediterranean basin	Africa	S. & SE. Asia	Australia & Oceania
Palembus spp.	●	X®	X			?	X	

Pest status: ● minor to ●●●● major pest
X: recorded
®: restricted distribution
?: status unclear

Palembus spp. are of tropical Asian origin. *P. dermestoides* is being spread to new areas on account of its use as a traditional medicine.

Smalleyed flour beetles (Genera: *Palorus* and *Coelopalorus*)

Coelopalorus foveicollis	
Palorus cerylonoides	
Palorus ficicola	
Palorus laesicollis	
Palorus genalis	
Palorus subdepressus	Depressed flour beetle
Palorus ratzeburgii	Smalleyed flour beetle

Summary

Feeding strategies	secondary pest, mould feeder
Commodities attacked	any dried material of animal and plant origin, but especially cereal grain and products
Distribution	*P. subdepressus* and *P. ratzeburgii* worldwide, others tropical
Economic importance	low
Eggs	laid amongst commodity
Larvae	elateriform, mobile, live amongst commodity
Adults	long lived, feed on commodity, can fly

Introduction

Members of the genera *Palorus* and *Coelopalorus* live under the bark of tree and in rotten wood. Several species are encountered as minor pests of stored grain, of which the most frequently encountered are *P. subdepressus* and *P. ratzeburgii*.

Identification

Palorus spp. and *Coelopalorus foveicollis* are flattened, parallel-sided, reddish-brown beetles. *Palorus* spp. (Figures 155–160) resemble a miniature dark glossy *Tribolium castaneum*. *C. foveicollis* (Figure 161) is about the same length as *T. castaneum*, more flattened in form and more glossy in apprearnce. When viewed from the side, the eyes of *Palorus* and *Coelopalorus* spp. are entire and

Figure 155 *Palorus ratzeburgii*, adult, head showing whole of hemisphere of eye visible from above

Figure 156 *Palorus ratzeburgii*, adult

Figure 157 *Palorus cerylonoides*, adult, head showing whole of hemisphere of eye visible from above

Figure 158 *Palorus cerylonoides*, adult showing head from side, eye round and undivided

Figure 159 *Palorus subdepressus*, adult, live

Figure 160 *Palorus subdepressus*, adult, head showing front of hemisphere of eye obscured by brow-like structure above eye

Figure 161 *Coleopalorus foveicollis*, adult showing depressions parallel with margin of thorax, elytra with vertical sides, ridge parallel with margin of elytra

round (Figure 158) and are not 'cut' almost in half by a sideways projection of the head capsule. Larvae are active, elateriform and light brown. Adults of the more important and distinctive species can be identified as below.

Key to *Palorus* and *Coelopalorus* species associated with stored products

1 Elytra with prominent lateral ridge and almost vertical lateral sides, pronotum with pair of depressions (foveae) parallel with side of pronotum, body very shiny, length of adult 3.6–4.3mm . . . . *C. foveicollis* (Figure 161)
Elytra without lateral ridge, body less shiny, length of adult 2.4–3.0 mm 2

2 Pronotum with pair of depressions (foveae), each running parallel with lateral edge of pronotum . . . . *P. laesicollis*
Pronotum without pair of depressions (foveae) 3

3 When viewed from above, whole of hemisphere of eye can be seen (Figure 155, 157) . . .
. . . . *P. ratzeburgii* (also *P. cerylonides*, *P. genalis* and *P. ficola*)
When viewed from above the front of hemisphere of eye is obscured from view from above by an expansion of the side of the head into a brow-like structure (Figure 160) . . .
. . . . *P. subdepressus*

Life cycle

Eggs are laid at random as females move through the food. The eggs are sticky when laid and as a result become coated with flour or other particles. Larvae are active and move through the food material. Pupation takes place within a cell made amongst the substrate. Adults are long lived (> 6 months), feed on the commodity and can fly. Females lay eggs over most of their life.

Physical limits and optimum rate of multiplication

Species	Conditions within which breeding takes place	Shortest development period, with optimum conditions	Maximum monthly rate of increase
Palorus ratzeburgii	17.5–40°C, r.h. > 20%	22 days at 32.5°C	
Palorus subdepressus	20–35°C, r.h. > 50%	36 days at 32.5°C	

P. ratzeburgii completes its development more rapidly and has a slightly wider temperature range and is more tolerant of low humidity than *P. subdepressus*.

Economic importance

Palorus spp. are minor pests of a wide range of stored produce. They are most often associated with residues or slightly damp grain, heated grain, mill machinery or grain that has already been attacked by other insects and has become contaminated by their faeces. They are often found in on-farm grain storage facilities, especially in tropical regions.

Type of damage and symptoms

Larvae and adults of *Palorus* and *Coelopalorus* spp. are general feeders, and damage is not readily identifiable as being specifically caused by this insect. Infestation by the Tenebrionidae leads to persistent disagreeable odours in the commodity. This is due to secretion of benzoquinones from abdominal glands.

Ecology

Under normal conditions of storage, *Palorus* and *Coelopalorus* spp. are likely to be out-competed by faster breeding species such as *Tribolium* spp. Presence of damaged grains, grain germ, moulds and frass of other insect species is beneficial to growth of populations of this insect. They are often found infesting commodities under conditions of tropical subsistence agriculture.

Monitoring

Palorus and *Coelopalorus* spp. are easily caught in pitfall type traps. Crevice traps are also effective and their efficacy can be improved with addition of food bait. A number of proprietary bait and trap systems are available which would be attractive to these pests.

Geographical distribution

Species	Pest status	USA & Canada	Central & South America	Europe & N.Asia	Mediterranean basin	Africa	S. & SE. Asia	Australia & Oceania
Coelopalorus foveicollis	●		X®			X	X	
Palorus cerylonoides	●					X	X	X®
Palorus ficicola	●		X®		X	X	X	

Species	Pest status	USA & Canada	Central & South America	Europe & N.Asia	Mediterranean basin	Africa	S. & SE. Asia	Australia & Oceania
Palorus laesicollis	●					X		
Palorus genalis	●						X	X
Palorus subdepressus	●●	X	X	X	X	X	X	X
Palorus ratzeburgii	●●	X	X	X	X	X	X	X

Pest status: ● minor to ●●●● major pest
X: recorded
®: restricted distribution

About six species of *Palorus* have been found associated with stored products. Only *P. ratzeburgii* and *P. subdepressus* are found worldwide with *P. ratzeburgii* being more frequently encountered in temperate regions than *P. subdepressus*. Other species have a more localised tropical distribution and are generally less important as pests.

Sitophagus hololeptoides

Summary

Feeding strategies	secondary pest, mould feeder
Commodities attacked	grain and grain products, especially maize
Distribution	Mexico, Central America, South America, west Africa
Economic importance	low
Eggs	laid amongst commodity
Larvae	elateriform, mobile, live amongst commodity
Adults	long lived, feed on commodity, can fly

Introduction

S. hololeptoides originates in Mexico and Central America as a scavenger and part-time predator. It is an occasional inhabitant of grain stored under conditions of tropical subsistence agriculture.

Identification

Adult beetles are 5–6 mm long, light reddish brown in colour (Figure 162). They are slightly oval in shape and highly flattened. From above, mandibles are clearly visible. When viewed from below, the pseudopleuron ('flange' at outer edge of elytra) is abruptly narrowed before tip of abdomen (Figure 163).

Life cycle

Eggs are laid at random as females move through the food media. Larvae are active and move through the food and pupation occurs within a cell made in the foodstuff. Adults are long lived, feed on the commodity and can fly.

Figure 162 *Sitophagus hololeptoides* adult, showing enlarged side of head and mandibles

Figure 163 *Sitophagus hololeptoides*, adult, elytra showing flange at margin of elytra abruptly narrows before tip of abdomen

Physical limits and optimum rate of multiplication

No specific data are available.

Economic importance

S. hololeptoides is a minor pest infesting farm stored grain especially maize stored under conditions of tropical subsistence agriculture.

Type of damage and symptoms

Larvae and adult *S. hololeptoides* are general feeders and damage is not readily identifiable as being specifically caused by this insect. Infestation can lead to persistent disagreeable odours in the commodity due to secretion of benzoquinones from abdominal glands.

Ecology

In nature, *S. hololeptoides* can be found living under bark of trees.

Monitoring

S. hololeptoides are easily caught in pitfall type traps. Crevice traps are also effective and their efficacy can be improved with addition of food bait.

Geographical distribution

Species	Pest status	USA & Canada	Central & South America	Europe & N.Asia	Mediterranean basin	Africa	S. & SE. Asia	Australia & Oceania
Sitophagus hololeptoides	●	X	X		X®			

Pest status: ● minor to ●●●● major pest
X: recorded
®: restricted distribution

S. hololeptoides has a restricted distribution, centred on tropical Central America. It has also been reported from west Africa.

Mealworms (Genus: *Tenebrio*)

Tenebrio molitor	Mealworm, Yellow mealworm
Tenebrio obscurus	Mealworm, Dark mealworm

Summary

Feeding strategies	secondary pest, scavenger
Commodities attacked	damp and mouldy plant material, animal and plant origin
Distribution	worldwide
Economic importance	low
Eggs	laid amongst commodity
Larvae	elateriform, mobile, live amongst commodity
Adults	long lived, feed on commodity, can fly

Introduction

Members of the genera *Tenebrio* live under the bark of tree and in rotten wood. Two species, *T. molitor* and *T. obscurus*, are minor pests, mainly of aged residues of stored grain. Larvae of *T. molitor* are widely reared and sold as a pet food.

Identification

Adult *Tenebrio* are 12 to 18 mm long, parallel-sided and flattened (Figures 164–167). They are reddish brown to black depending on age of the specimen. Larvae are elateriform, brown and about 25–30 mm long when fully grown. Adults of the two species can be distinguished as below.

3mm

Far left: Figure 164 *Tenebrio molitor*, adult

Left: Figure 165 *Tenebrio molitor*, adult, thorax with glossy surface and pits on surface not touching each other

3mm

Far left: Figure 166 *Tenebrio obscurus*, adult

Left: Figure 167 *Tenebrio obscurus*, adult, thorax with matt surface and pits on surface touching each other

Identification of *Tenebrio* species associated with stored products

Punctures (pits) on pronotum of adult spread out and not touching each other (Figure 165), adult has slightly glossy appearance (Figure 164) .... *T. molitor*

Punctures (pits) on pronotum of adult dense and touching each other (Figure 167), adult has matt appearance (Figure 166) .... *T. obscurus*

Physical limits and optimum rate of multiplication

Species	Conditions within which breeding takes place	Shortest development period, with optimum conditions	Maximum monthly rate of increase
Tenebrio molitor	14-30°C, r.h. > 30%	120 days at 25–27°C	

Population development of these insects is slow and very variable in length. Only one to two generations a year are typically produced. Adult beetles can live for one to two years. Larvae are highly resistant to cold (three weeks at −12°C) and low humidity.

Economic importance

Tenebio spp. are minor pests and scavengers of a wide range of cereals and cereal products, especially if damp and in poor condition, such as aged residues. *T. molitor* is widely reared as a pet food.

Type of damage and symptoms

Larvae and adult *Tenebio* spp. are general feeders and damage is not readily identifiable as being specifically caused by this insect. Infestation can lead to persistent disagreeable odours in the commodity due to secretion of benzoquinones from abdominal glands.

Ecology

In mills and grain storage, large populations of these insects usually only occur under conditions of poor storage and hygiene. Adults and larvae of this species readily prey on other insects present.

Monitoring

Tenebio spp. are easily caught in pitfall type traps. Crevice traps are also effective and their efficacy can be improved with addition of food bait.

Geographical distribution

Species	Pest status	USA & Canada	Central & South America	Europe & N.Asia	Mediterranean basin	Africa	S. & SE. Asia	Australia & Oceania
Tenebrio molitor	●	X	X	X	X	X	X	X
Tenebrio obscurus	●	X	X	X	X	X	X	X

Pest status: ● minor to ●●●● major pest
X: recorded

Tenebrio spp. are cosmopolitan but are most frequently encountered in temperate regions. They are unable to breed in hot tropical climates.

Flour beetles (Genus: *Tribolium*)

Tribolium audax	American black flour beetle
Tribolium castaneum	Rust red flour beetle, Red flour beetle
Tribolium confusum	Confused flour beetle
Tribolium destructor	False black flour beetle
Tribolium madens	Black flour beetle

Summary

Feeding strategy	secondary pest (primary pest?)
Commodities attacked	dried material of animal and plant origin, but especially cereal grain and products
Distribution	*T. confusum* and *T. castaneum* worldwide, others have restricted distributions
Economic importance	*T. confusum* and *T. castaneum* high, others variable
Eggs	laid amongst commodity
Larvae	elateriform, mobile, live amongst commodity
Adults	long lived, feed on commodity, some species fly

Introduction

The genus *Tribolium* consists of some 30 species. In nature they live under the bark of trees and in animal nests where they feed on other insects and detritus of animal and plant origin. A number of species are associated with stored products. Two of these, *T. castaneum* and *T. confusum*, are amongst the most important pests of stored produce worldwide.

Identification

Worldwide some five *Tribolium* species are frequently found associated with stored products (Figures 168–184). These can be identified by using the key below. By far the most frequently encountered are *T. castaneum* and *T. confusum*. In Europe, North America and parts of temperate

Figure 168 *Tribolium audax*, adult

Figure 169 *Tribolium audax*, adult, head from above, pits on surface between eyes running together, terminal segment of antennae somewhat square

Figure 170 *Tribolium audax*, adult, underside head, eye does not extend as far as lateral margin of maxillary fossa (Lm)

Figure 171 *Tribolium castaneum*, adult, live

Figure 172 *Tribolium castaneum*, adult, head side view, showing division of eye, eye two facets wide at narrowest point

Figure 173 *Tribolium castaneum*, adult, head underside, gap between eyes relatively narrow, obvious three segmented antennal club, process between front legs shaped like axe head

Figure 174 *Tribolium castaneum*, adult, head from above, obvious brow over eye, pits on surface in centre of thorax small

Figure 175 *Tribolium castaneum*, adult, infestation

Figure 176 *Tribolium castaneum*, larva

Figure 177 *Tribolium confusum*, adult, head underside, gap between eyes relatively wide, segments of antennae gradually get wider towards tip, process between front legs shaped like axe head

Figure 178 *Tribolium destructor*, adult

Figure 179 *Tribolium destructor*, adult, underside, showing somewhat pointed end of final antennal segment

Figure 180 *Tribolium destructor*, adult, head from above, pits on surface in centre of thorax large

Figure 181 *Tribolium madens*, adult

Figure 182 *Tribolium madens*, adult, underside of head, eye extends as far as lateral margin of maxillary fossa (Lm)

Figure 183 *Tribolium madens*, adult, side view of head showing division of eye, eye more than four facets wide at narrowest point

Figure 184 *Tribolium madens*, adult, head from above, tip of antennae square, pits on surface between eyes do not meet

Asia, other species listed may also be present. These species are characterised by being bigger and/or darker than *T. castaneum* and *T. confusum*. Larvae of *Tribolium* are leathery, elongate and elateriform. It is possible but difficult to identify larvae and even pupae to species – see keys in Sokoloff (1972) and Spilman (1987b).

Key to adult *Tribolium* species associated with stored products

1   Tip of last segment of antennae is somewhat pointed (Figures 177, 179). When viewed from side, head has ridge (carina) immediately above eye, middle of eye at narrowest point 1 to 2 facets wide (Figure 172). Colour – reddish to dark brown 2
      Tip of last segment of antennae is somewhat square (Figures 170, 182). When viewed from side, head without ridge (carina) immediately above eye, middle of eye at narrowest point 4+ facets wide (Figure 183). Colour – black or brownish-black 4

2   Colour – reddish brown, length 2.6–4.4 mm, punctures in centre of pronotum small (Figure 174) 3
      Colour – dark brown to black, length 4.5–5.7 mm, punctures in centre of pronotum large (Figure 180) . . . . *T. destructor* (Figure 178)

3   When viewed from underneath gap between eye is relatively narrow – about 33% of
    head width. Last three segments of antennae form distinctive club (Figure 173) ......
    .... *T. castaneum* (Figures 171, 175)
    When viewed from underneath gap between eye is relatively wide – about 50% of head
    width. Last three segments of antennae get gradually wider and do not form distinctive
    club .... *T. confusum* (Figure 177)

4   From above, many punctures meet and run into each other between eyes (Figure 169).
    When viewed from underneath, eye does not extend as far as lateral angle of maxillary
    fossa (Figure 170), length 2.8–4.5 mm .... *T. audax* (Figure 168)
    From above, punctures do not meet or run into each other between eyes (Figure 184).
    When viewed from underneath, eye extends as far as lateral angle of maxillary fossa
    (Figure 182), body length 3.9–5.1 mm .... *T. madens* (Figure 181)

Life cycle

Data provided is for *T. castaneum*. Eggs are laid at random amongst the commodity. They are
sticky and soon become coated with flour or other particles. Females may lay up to 1000 eggs
over most of their lifetime. Larvae are elateriform and are active and move through the food
(Figure 176). They feed on the commodity and on other insects that are small enough to subdue.
Cannibalism amongst larvae and adults is common. Full grown larvae of *T. castaneum* are about
10 mm long. Pupae are naked and are also found amongst the food. Egg and pupal stages are
relatively brief and more than 60% of development time is spent as larvae. Adults are long lived,
up to two to three years under temperate conditions.

Physical limits and optimum rate of multiplication

Species	Conditions within which breeding takes place	Shortest development period, with optimum conditions	Maximum monthly rate of increase
Tribolium castaneum	22–40°C, r.h. > 1%	20 days at 35–37.5°C, > 70% r.h.	70
Tribolium confusum	19–37.5°C, r.h. > 1%	25 days at 32.5°C, 70% r.h.	60
Tribolium destructor	Max 30°C, r.h. > 10%	44 days at 28°C, 75% r.h.	
Tribolium madens	20–35°C, r.h. > 10%	35 days at 35°C, 70% r.h.	

Under optimal conditions, growth of *T. castaneum* and *T. confusum* populations are among the
most rapid achieved by insect pests of stored products. *T. confusum* is able to breed under slightly
cooler conditions than *T. castaneum*, which may explain why *T. confusum* is more common in
temperate regions. Both species are, however, remarkably tolerant of low humidity. Development
periods for other species are longer. *T. destructor* appears intolerant of temperatures higher than
30°C which may account for this species being limited to areas with a cool climate.

Economic importance

Worldwide *T. castaneum* and *T. confusum* are major and frequently encountered pests of stored
products. They attack virtually any dried material of animal or plant origin but are especially

important as pests of cereals and cereal products and are major pests of mills. *T. castaneum* occurs in both grain stores and mills, however, *T. confusum* is more often found in mills. *T. audax, T. destructor* and *T. madens* are also pests of cereal grain and grain products and can be locally important in regions where they are found. For example in southern Canada, *T. destructor* can be as important a pest as *T. castaneum* and *T. confusum.*

Type of damage and symptoms

Larvae and adult *Tribolium* spp. are general feeders and damage is not readily identifiable as being specifically caused by this insect. Infestation can lead to persistent disagreeable odours in the commodity due to secretion of benzoquinones from abdominal glands.

Ecology

Tribolium spp. have been long associated with stored products, as evidenced by the remains of *T. confusum* found in a sealed jar in an Egyptian tomb of *circa* 2500 BC. Their size, wide food range, long lifespan and ability to cope with very low humidities make *Tribolium* spp. highly successful pests of stored produce.

Especially under tropical conditions, *T. castaneum* is a coloniser species which is frequently the first pest to infest a stored commodity. It is also a ubiquitous pest of grain handling and transportation systems and is frequently found by quarantine officials infesting residues in empty shipping containers used in international trade.

Competition between *T. confusum* and *T. castaneum* is complex and not fully understood. Mixed populations only appear to exist at low density and as competition increases one or other species tends to dominate. One major difference between these species is their ability to disperse. *T. castaneum* flies readily under warm conditions. In contrast, *T. confusum* does not appear to fly so is much more reliant upon movement by human intervention. *T. confusum* may therefore persist better in places such as mills, where long-term populations can develop in machinery and residues whereas *T. castaneum* is a better coloniser species. *Tribolium* spp. also readily prey on other insects. When present in large numbers they may have a significant effect on other pests and potential competitors.

Different *Tribolium* species are at different stages of development as storage pests. At one extreme are *T. confusum* and *T. castaneum* and at the other extreme are species still only found in natural habitats. In between are species that are either locally distributed storage species such as *T. audax, T. destructor* and *T. madens*, or species which are only occasionally associated with stored products such as *Tribolium brevicorne* and *Tribolium freemanii*. It is possible that over time some of these additional species may become new or more serious storage pests.

Monitoring

Tribolium spp. are easily caught in pitfall type traps. Crevice traps are also effective and their efficacy can be improved with addition of food bait. A number of proprietary bait and trap systems are available which would be attractive to these pests. Trapping systems baited with synthetic aggregation pheromones are commercially available for *Tribolium* spp.

Geographical distribution

Species	Pest status	USA & Canada	Central & South America	Europe & N.Asia	Mediterranean basin	Africa	S. & SE. Asia	Australia & Oceania
Tribolium audax	••	X						
Tribolium castaneum	••••	X	X	X	X	X	X	X
Tribolium confusum	••••	X	X	X	X	X	X	X
Tribolium destructor	•••	X	X	X	X			
Tribolium madens	••	X	X	X				

Pest status: • minor to •••• major pest
X: recorded

T. castaneum and *T. confusum* are found worldwide. Under tropical conditions, *T. castaneum* dominates. In warm temperate and Mediterranean climates, both species are encountered, *T. castaneum* tends to be associated with cereal grain and grain storage whereas *T. confusum* largely replaces *T. castaneum* in mills. Cooler climates tend to favour *T. confusum* in any habitat.

T. audax, T. destructor and *T. madens* are largely restricted to temperate regions or cool areas in otherwise tropical countries. These species are of interest to quarantine services as they are currently absent from many areas in which they would likely survive, for example Australia. The ability of these species to spread is illustrated by the establishment of *T. madens* in the USA and Canada following its initial detection in the state of Kentucky in 1977 and Canada in 1979.

Trogossitid beetles
(Family: Trogossitidae)

Tenebroides mauritanicus	Cadelle

Summary

Feeding strategies	secondary pest, scavenger
Commodities attacked	stored products of vegetable origin
Distribution	worldwide
Economic importance	low
Eggs	laid in amongst commodity
Larvae	campodeiform, mobile, external feeders
Adults	long lived, feed on commodity

Introduction

In nature, trogossitid beetles are found living under the bark of trees, often as predators and scavengers. One species, *Tenebroides mauritanicus*, is associated with stored products.

Identification

Adult *T. mauritanicus* are distinctive flattened, parallel-sided, glossy black beetles, 6–11 mm long (Figure 185). The head and pronotum are relatively large when compared to the rest of the body. The prothorax and elytra are distinctively separated by a visible constriction or 'waist'. Larvae are elongate, flattened and white in colour, the thorax and last segment of abdomen marked with dark coloured areas.

Figure 185 *Tenebroides mauritanicus*, adult, thorax and abdomen separated by waist

Life cycle

T. mauritanicus lays clusters of 10–50 eggs into cracks and crevices. About 1000 eggs can be laid by a female in her lifetime, which can be as long as several years. Larvae feed both on the stored commodity and on other insects present. When feeding on grain, larvae preferentially attack the germ. Pupation takes place in a cell hollowed out in a solid substrate. Both adult and larvae of *T. mauritanicus* are cold-tolerant and can survive periods below 0°C. Development from egg to adult can take place in as little as 70 days.

Economic importance

T. mauritanicus is a pest of a wide range of stored products, especially cereals, oilseeds and their products in mills and processing facilities. Larvae can cause serious damage by burrowing into wooden structures and even plastered walls.

Ecology

Burrowing by *T. mauritanicus* larvae into foodstuffs and storage structures provides harbourage for other insect pests. *T. mauritanicus* is rarely seen in large numbers, except under conditions of long established poor hygiene and pest control.

Type of damage and symptoms

Larvae and adults are general feeders, and damage is not readily identifiable as being specifically caused by this insect. Larvae of *T. mauritanicus* often burrow into wood structures and plastered walls.

Monitoring

T. mauritanicus is easily caught in pitfall type traps. Crevice traps are also effective and their efficacy can be improved with addition of food bait.

Geographical distribution

Species	Pest status	USA & Canada	Central & South America	Europe & N.Asia	Mediterranean basin	Africa	S. & SE. Asia	Australia & Oceania
Tenebroides mauritanicus	●●	X	X	X	X	X	X	X

Pest status: ● minor to ●●●● major pest   X: recorded

T. mauritanicus is found worldwide, including temperate regions.

References

Aitken (1975), Arbogast (1991), Haines (1991).

Moths and Butterflies (Order: Lepidoptera)

The Lepidoptera, or butterflies and moths, are characterised by adult insects having wings covered with coloured scales. Where present in the adult, mouthparts mostly take the form of a tube or proboscis (see below) through which the insect takes liquid food such as nectar. Metamorphosis is complete: eggs hatch into larvae, the feeding and growing stage, which then pupate to become pupae, a transition stage from which the adult emerges.

Larvae of Lepidoptera have biting mouthparts and feed mainly on plant material, and a few species feed on material of animal origin such as wool and feathers. Those of interest as pests of stored products belong to the relatively small number of families that feed on dead and dried plant and animal materials. Of these, the most important are members of the families Pyralidae and Gelechiidae. In addition, members of the Oecophoridae and Tineidae are present, mainly as scavengers.

To identify moths of stored products a basic understanding of their structure is needed. Terms illustrated below (Figures G–J) are widely used in the keys and descriptions in this chapter.

Figure G. Major body components of an adult moth

Moths use both fore and hind wings for flight, unlike beetles, which use only their hind wings. The wings of moths are covered with coloured scales which provide wing colour and pattern. In moths associated with stored products, wings rapidly get rubbed and damaged due to contact with infested commodities.

In contrast to beetles, the mouthparts of adult moths are highly modified to suck liquid food such as nectar. Externally, two structures are obvious: the labial palps and the proboscis (Figure H). The labial palps have a sensory function. Their length and physical form are important in the identification of storage moths. Between the palps is the long tube-like sucking mouth or proboscis. This structure is unique to moths and butterflies. When not in use, the proboscis is stored away coiled like a flat spring.

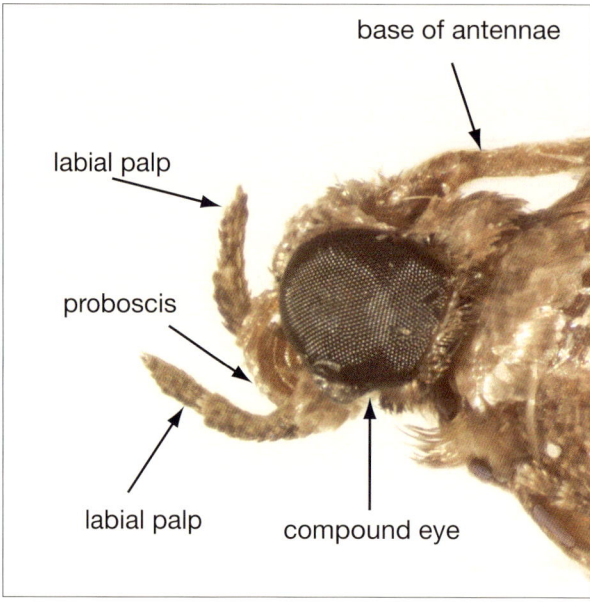

Figure H. Structure of the head of a moth

Adult moths, like all insects, have only three pairs of jointed legs (true legs), which are attached to segments of the thorax. Many moth larvae have additional structures known as pro-legs, which emerge from abdominal segments (Figure I). They are not jointed (unlike true legs) and assist in the locomotion of the larva's soft long tube-like body. The pro-legs are lost on pupation and development to adulthood.

All insects respire via branching tubes known as trachea. These connect to the insect's outer surface via small valve-like openings known as spiracles, which are present on either side of each segment of the thorax and abdomen. Spiracles are easy to see on the bodies of fully grown moth larvae (Figure J) and are of use in identification.

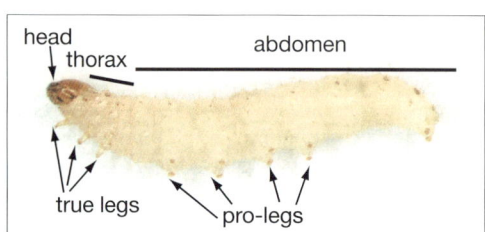

Figure I. Major body components of a moth larva

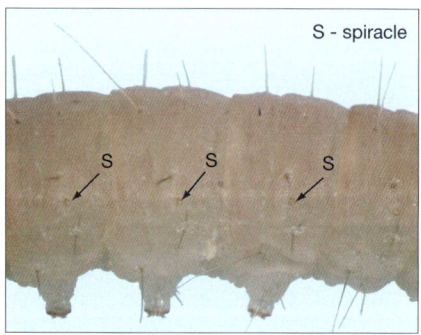

Figure J. Detail of abdomen of moth larva

Key to major families of moths associated with stored products

1   Labial palps long and strongly curved upwards (Figure 189) 2
     Labial palps not strongly curved upwards (Figure 194) 3

2   Fore wing 6–10 mm in length, mottled or bloched; fore and hind wings fringed with hairs and rounded at apex (Figures 186, 187) .... Oecophoridae (Pages 123–125)
     Fore wing 6 mm or less in length, light brown, small black dot in centre of wing; wings heavily fringed with fine hairs. The front wing is tapered to apex, rear wing shaped like old fashioned 'fingerboard' road sign (Figure 188) .... Gelechiidae (Pages 125–128)

3   Fore wings marked with dark brown/black blotches on light background (Figure 208), head covered in rough erect scales (Figure 207), labial palps short and not curved upwards .... Tineidae (Pages 134–136)
     Fore wings grey (Figures 192, 195), bicoloured (Figure 202) or patterned (Figure 206), head covered in smooth scales (193, 203); Labial palps variable in length (Figures 193, 194), pointing forwards (Figure 203) or downwards (Figure 194) ....
     .... Pyralidae (Pages 128–134)

Oecophorid moths
(Family: Oecophoridae)

Endrosis sarcitrella	White-shouldered house moth
Hofmannophila pseudospretella	Brown house moth

Summary

Feeding strategy	scavenger
Commodities attacked	wide variety of plant material including grain and grain products
Distribution	worldwide, especially temperate regions
Economic importance	low
Eggs	laid on cracks and crevices
Larvae	active, external feeders
Adults	short lived, do not feed on commodity, can fly

Introduction

Moths of this family have very long strongly curved labial palps. Larvae of this family are concealed feeders on dried and growing plant material in a wide range of environments. Two species, *Endrosis sarcitrella* and *Hofmannophila pseudospretella*, are associated with stored products, mainly as scavengers.

Identification

Fresh adult *E. sarcitrella* are distinctive in appearance (Figure 186), with the head and thorax clothed in white scales. The fore wing (length 6–10 mm) is cream heavily specked with dark brown scales. Adult *H. pseudospretella* are larger (fore wing 6–12 mm in length). The head and thorax is brown and the fore wings a mottled brown (Figure 187). *H. pseudospretella* may be confused with some pyralid moths on account of its general non-descript appearance, but can be distinguished from them by its strongly upwardly curved upwards labial palps.

Figure 186 *Endrosis sarcitrella*, adult, obvious wing patterning, fringed hind wings, head and front of thorax covered in white scales

Figure 187 *Hofmannophila pseudospretella*, adult, fore wings mottled brown

Life cycle

Adults lay eggs in crevices, larvae move through food material feeding as they go. Pupation may occur in or near affected food material.

Physical limits and optimum rate of multiplication

Species	Conditions within which breeding takes place	Shortest development period, with optimum conditions	Maximum monthly rate of increase
Endrosis sarcitrella	10–29°C, r.h. > 70%	62 days at 29°C, 90% r.h. Up to 4 generations a year	
Hofmannophila pseudospretella	10–29°C, r.h. > 80%	One generation a year	

Development of these species is relatively slow when compared to the more important pyralid pest species. *E. sarcitrella* will breed continuously if conditions are suitable, whereas *H. pseudospretella* has one generation a year, over-wintering as a larva. The low minimum temperature at which these moths can breed permits them to be pests in cool temperate climates.

Economic importance

Both species are scavengers. Presence of large numbers of these insects suggests presence of aged residues or a history of poor hygiene. *E. sarcitrella* is more often associated with grain stores. *H. pseudospretella* is more often found in domestic situations and has been reported attacking carpets (larvae of this species are capable of digesting wool) and corks in wine bottles.

Type of damage and symptoms

Larvae produce silk webbing as they feed and pupate. Irregular holes may be bitten into attacked material.

Ecology

Both species are frequent inhabitants of birds' nests, a probable natural habitat and source of infestation. These species often occur together.

Monitoring

No specific trapping systems appear to be available for these species.

Geographical distribution

Species	Pest status	USA & Canada	Central & South America	Europe & N.Asia	Mediterranean basin	Africa	S. & SE. Asia	Australia & Oceania
Endrosis sarcitrella	●	X	X*	X	X	X*	X*	X
Hofmannophila pseudospretella	●	X	X*	X	X*	X*	X	

Pest status: ● minor to ●●●● major pest
X: recorded
*: Temperate regions only

Both species are found in temperate regions, and both appear unable to breed in hot tropical climates.

References

Aitken (1984), Cox and Bell (1991), Ferguson (1987), Mound (1989), Solis (1999) and Weismann (1987).

Gelechiid moths
(Family: Gelechiidae)

Sitotroga cerealella	Angoumois grain moth

Summary

Feeding strategy	primary pest
Commodities attacked	grain
Distribution	warm temperate–tropical
Economic importance	high
Eggs	laid on grain
Larvae	immobile when mature, internal feeders
Adults	short lived, do not feed on commodity, fly readily

Introduction

Moths of this family have very long strongly curved labial palps. Larvae of this family are concealed feeders on dried and growing plant material in a wide range of environments. Only one species, *Sitotroga cerealella*, is an important pest of whole cereal grains.

Identification

S. cerealella is unlikely to be confused with other moths of stored produce on account of its general appearance, size, coloration when fresh, and lifestyle (Figures 188–191).

Sitotroga adults are smaller than other commonly encountered moth pests of stored products. When fresh, the wings are pale greyish brown, 5–6 mm long (Figure 188). A single small black spot is present in fresh specimens roughly central on the fore wing two-thirds from the base. Wings are heavily fringed with fine hairs. The front wing is tapered to apex and when unfurled the rear wing is shaped like old fashioned 'fingerboard' road sign. Labial palps are long and curled upwards (Figure 189).

Larvae are not usually seen as they live concealed inside grains (Figure 191). True legs are reduced in size and pro-legs on the abdomen are vestigial. As previously mentioned, larvae of *S. cerealella* can be confused with some beetle larvae. For a description of differences, see page 17.

Figure 188 *Sitotroga cerealella*, adult, wings heavily fringed, fore wing tapered, tip of hind wing constricted, like old fashioned 'fingerboard' road sign

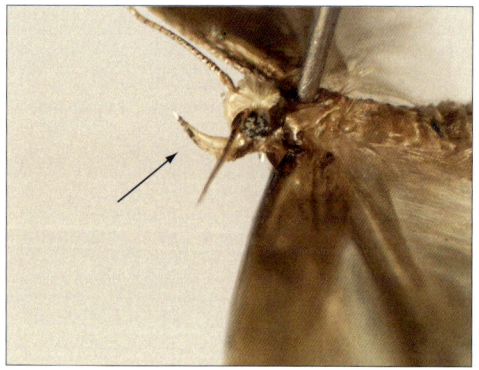

Figure 189 *Sitotroga cerealella*, adult, head showing labial palps long and strongly curved upwards

Figure 190 *Sitotroga cerealella* adult, live in wheat

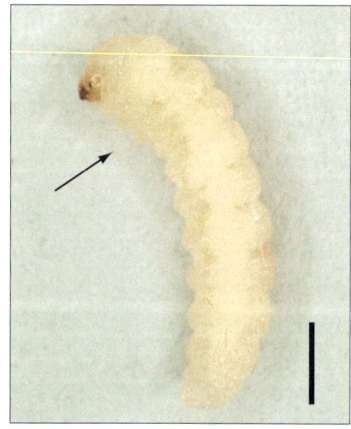

Figure 191 *Sitotroga cerealella*, larva, effective absence of pro-legs on abdomen

Life cycle

Eggs are laid on the outside of grains in cracks and crevices, either singly or in groups and as they develop they become red in colour. Up to 150 eggs are laid by each female. Eggs hatch in four to six days under optimal conditions. Newly hatched larvae burrow into a grain inside which they remain until emergence as an adult. Larvae excavate a cavity inside the grain as they feed. Prior to pupation, larvae make an emergence hole which they cover with a thin, tough envelope of silk ('emergence window'). After about seven days, adults emerge, escaping through the 'window'. Adults are short lived, do not feed and are most active at dusk and during the night.

Physical limits and optimum rate of multiplication

Species	Conditions within which breeding takes place	Shortest development period, with optimum conditions	Maximum monthly rate of increase
Sitotroga cerealella	16–35°C, r.h. > 30%	30 days at 30°C, 75% r.h.	50

S. cerealella is cold hardy and can survive winter in unheated premises in temperate areas.

Economic importance

S. cerealella is a pest of whole cereal grain, especially of barley, maize, millet, rice, sorghum and wheat. Infestation often begins in the field as the grain matures, especially under tropical conditions. It is most serious a pest of bagged and traditionally stored produce and commodities such as maize and sorghum, which are often left in the field or in the open to dry. It is less serious as a pest of bulk-stored grain where infestations are usually restricted to surface layers.

Type of damage and symptoms

Feeding by larvae produces large cavities within the infested grain. When adults emerge the pupal case is often characteristically left sticking half out of a neat round emergence hole. Unlike other moths, *Sitotroga* does not bind grains etc. with silk. *Sitotroga* infestations produce a lot of heat and moisture in otherwise dry grain. This can encourage mould growth and attract other insect species.

Ecology

Infestations of *S. cerealella* frequently begin in the field before harvest. Large populations can build up if crop is left in the field to dry. Sorghum with its loose seed head is especially at risk from this pest. In heavy mixed populations, *S. cerealella* may eventually be suppressed by internal feeding beetles such as *Sitophilus* species and *Rhyzopertha dominica*.

Monitoring

Trap systems baited with a synthetic pheromone are commercially available for this species.

Geographical distribution

Species	Pest status	USA & Canada	Central & South America	Europe & N.Asia	Mediterranean basin	Africa	S. & SE. Asia	Australia & Oceania
Sitotroga cerealella	●●●	X	X	X	X	X	X	X

Pest status: ● minor to ●●●● major pest
X: recorded

S. cerealella is cosmopolitan and is widely distributed in mild and warm temperate to tropical regions. Heavy infestations have occurred in areas such as central and southern Europe and southern USA.

References

Aitken (1984), Cox and Bell (1991), Haines (1974, 1981, 1991), Ferguson (1987), Mound (1989) Solis (1999) and Weismann (1987).

Pyralid moths
(Family: Pyralidae)

Cadra cautella [†]	Almond moth, Warehouse moth
Cadra figulilella [†]	Raisin moth
Corcyra cephalonica	Rice moth
Ephestia calidella	Carob moth, Dried fruit moth
Ephestia elutella	Tobacco moth
Ephestia kuehniella *	Mediterranean flour moth
Plodia interpunctella	Indian meal moth
Pyralis farinalis	Meal moth

([†] sometimes included in genus *Ephestia*)

(* also known as *Anagasta kuehniella*)

Summary

Feeding strategies	secondary pest, scavenger
Commodities attacked	grain and grain products, oilseeds, nuts, herbs and spices, dried fruit, tobacco
Distribution	worldwide
Economic importance	high – especially on processed foods
Eggs	laid in amongst commodity
Larvae	mobile, external feeders, produce large quantities of silk
Adults	short lived, do not feed on commodity, fly readily

Introduction

Pyralid moths occur worldwide and feed on dried and growing plant material in a wide range of environments. Members of several genera, notably *Cadra*, *Corcyra*, *Ephestia* and *Plodia* are important pests of a wide range of stored products, especially milled, processed and manufactured produce. *Pyralis farinalis* is a minor pest of mills and storage residues.

Identification

Fresh adults of *P. interpunctella* and *P. farinalis* are distinctive. *Cadra* and *Ephestia* species are superficially similar. Members of these genera are best identified by dissection and examination of genitalia, which is easiest in male specimens (see references). Larvae of pyralid moths can be identified to genus easily.

Genus: *Corcyra* (Figures 192–194)

Adult
- Fore wings (when fresh) grey with no markings (Figure 192). Fore wing 8–13 mm in length. Males much smaller than females. At rest, tip of fore wings appear more tapered than in *Cadra* and *Ephestia* spp.

  Labial palps: female – long, curved downwards (Figure 194); male – short, hidden by scales (Figure 193).

Larva
- Full grown about 15–20 mm long, white in colour, rim of abdominal spiracles obviously thickened on one (rear) side.

Figure 192 *Corcyra cephalonica*, adult, tapered fore wing, with lack of pattern

Figure 193 *Corcyra cephalonica*, head of male, labial palps short

Figure 194 *Corcyra cephalonica*, head of female, labial palps long and pointing forward

---

Genera: *Cadra* (Figures 195–197) and *Ephestia* (Figures 198–200)

Adult
- Fore wings (when fresh) grey with vague darker markings (Figures 195, 198, 199). Fore wing 7–14 mm in length. Labial palps curved upwards (Figure 196).

Larva
- Full grown about 15–20 mm long, white to pink in colour, marked with black spots (each at base of a hair), rim of abdominal spiracles evenly thickened (Figure 197).

Figure 195 *Cadra cautella*, adult, live

Figure 196 *Cadra cautella*, adult, head showing labial palps curved upwards

Figure 197 *Cadra cautella*, larva, live, hairs on body each emerging from small dark spot

Figure 198 *Ephestia elutella*, adult

Figure 199 *Ephestia kuehniella*, adult

Figure 200 *Ephestia kuehniella*, larva, hairs on body each emerging from small dark spot

Plodia interpunctella (Figures 201–205)

Adult
- Fore wings (when fresh) are bi-coloured cream and reddish brown. Fore wing about 7–9 mm in length (Figure 201–202).
- Labial palps point forwards (Figure 203).

Larva
- Full grown about 15 mm long, creamy white in colour, not marked with black spots, rim of abdominal spiracles evenly thickened (Figures 204–205).

Figure 201 *Plodia interpunctella*, adult, live

Figure 202 *Plodia interpunctella*, adult, bicoloured pattern of fore wing

Figure 203 *Plodia interpunctella*, adult, head showing labial palps pointing forwards

Figure 204 *Plodia interpunctella*, larva, no pigmented spots at base of hairs

Figure 205 *Plodia interpunctella*, larva, detail of abdomen

Pyralis farinalis (Figure 206)

Adult
- Wings richly patterned. Fore and hind wing each with wavy while line across them, delineating basal (close to body) and apical (towards wingtip) purplish-brown regions. Central band pale brown (Figure 206).
- Fore wing broad and > 10 mm in length

Larva
- Full grown 20–25mm long, creamy while in colour, not marked with black spot

5 mm

Figure 206 *Pyralis farinalis*, adult, pattern of fore wing

To identify specimens further to species, see Haines (1991), Ferguson (1987), Mound (1989), Solis (1999) and Weismann (1987).

Life cycle

Eggs are laid at random over the food material. Eggs of *Cadra* and *Ephestia* are laid loose, those of *Plodia* are sticky and adhere to the food substrate. Some 150 to 200 eggs are produced during the short life of the adult. Eggs hatch in about three days at 30°C (*C. cautella*). Egg laying occurs at dusk or dawn when the moths are most active. Adult moths are able to follow odour trails from attractive foods and may fly in from some distance. Newly hatched larvae wander in search of food, and will readily enter minute imperfections in product packages. Larvae burrow into food bulks, lining and reinforcing their tunnels with silk as they go. In temperate regions, a combination of declining temperature and shortening day-length may cause mature larvae of *Cadra*, *Ephestia* and *Plodia* to enter diapause. In this state they can suspend development for months, until conditions improve.

When mature, larvae tend to wander in search of a pupation site, during which time they frequently become obvious to workers having emerged from concealed feeding places. Prior to pupation, larvae spin a cocoon, less dense than one made for diapause. The pupal stage of *C. cautella* lasts about seven days at 30°C. Upon emergence, females emit a pheromone to which males are attracted and mating occurs shortly after emergence. Flight activity is highest at dusk with a smaller peak at dawn. During daylight hours moths usually rest. Moths are short lived and do not feed, but longevity and egg production are improved if females have access to water.

Physical limits and optimum rate of multiplication

Species	Conditions within which breeding takes place	Shortest development period, with optimum conditions	Maximum monthly rate of increase
Cadra cautella	17–37°C, r.h. > 20%	26 days at 30°C, 75% r.h.	60
Cadra figulilella	15–35°C	27 days at 30°C, 70–90 % r.h.	20
Corcyra cephalonica	17–35°C, r.h. > 20%	27 days at 30°C, 75% r.h.	10
Ephestia elutella	10–30°C, r.h. > 25%	35 to 42 days at 25°C, 70% r.h.	15
Ephestia kuehniella	12–30°C, r.h. > 0%	40 days at 25°C, 75% r.h.	50
Ephestia calidella	15–35°C, r.h. > 20%	23 days at 30°C, 70% r.h.	
Plodia interpunctella	15–35°C, 25–90% r.h.	30 days at 30°C, 75% r.h.	60

C. cautella, *C. cephalonica* and *P. interpunctella* breed most rapidly under tropical conditions. Both *E. elutella* and *E. kuehniella* are able breed at lower temperatures than other major pest species and caterpillars in diapause are sufficiently cold-hardy to survive winter unaided in cool temperate areas. Both *E. elutella* and *E. kuehniella* do not cope well with high temperatures, as above 30°C they become infertile. All species are tolerant of low humidities and foodstuffs with low moisture contents.

Economic importance

While these moths attack a wide range of dried food materials they are especially important pests of high value processed commodities which include cereal products, oilseeds, cocoa, chocolate, spices, tobacco, nuts and dried fruit. These moths are frequently encountered in mills, food processing plants, warehouses, shops and domestic premises. Populations usually survive in residues in structures and machinery.

In warm temperate to tropical regions, *C. cautella* is a major pest of mills and food processing plants. It is sometimes replaced in tropical areas by *C. cephalonica*. In temperate regions, *C. cautella* tends to be replaced by *E. kuehniella* and/or *E. elutella*. As an example, in southern Australia, *C. cautella* and *E. kuehniella* can both be found infesting cereal processing plants with the former being mostly confined to warm or heated areas and the latter found in unheated facilities or areas exposed to ambient conditions. As well as being a pest of mills, *E. elutella* is also known as a pest of cured tobacco. *E. figulilella* and *E. calidella* are relatively minor pests and typically infest dry and drying fruit – infestations can begin on the vine before harvest. *P. farinalis* is a minor pest of mills, usually associated with aged residues, sweepings and composting vegetable matter. *C. cautella* and *E. kuehniella* will also infest bulk and bag stored grain. Infestations of bulk grain tend to be restricted to surface layers.

In warm-temperate and sub-tropical climates such as found in Australia and the southern USA, *P. interpunctella* is by far the most important storage pest of processed, packaged and manufactured food and confectionary products in manufacturing, distribution, retail and domestic environments. As a result, *P. interpunctella* ranks as one of the most important storage pest in terms of its economic impact. In these regions it is also the storage pest that is most likely to be encountered in domestic and retail environments.

Type of damage and symptoms

When feeding on whole grain, larvae preferentially feed on the germ and bran layer. As they feed, caterpillars produce large quantities of silk which binds together and fouls the infested commodity. Large larvae may bore into neighbouring packages, especially when they are packed together close inside a shipping carton. Silk produced can block machinery and act as a harbourage for other insect pests. Infested food becomes contaminated with silk, frass, cast skins, pupal cases and dead moths. Adult moths do not feed on the commodity, having only sucking mouthparts.

Ecology

In addition to the storage environment, the pest species described above have been found in a wide range of non-storage habitats, including rotting plant material, wasp nests, bee hives, under bark and in rotting wood.

When in diapause, larvae can become much more tolerant to fumigants and pesticides than when they are active. The ability to enter diapause varies, with strains and species from cold climates having the greatest capability.

Monitoring

Trap systems baited with a synthetic sex pheromone are commercially available for *Cadra*, *Ephestia* and *Plodia* species. These traps are highly effective at attracting male moths and are in widespread use. While specific pheromone blends are marketed for target species there is considerable cross-sensitivity between these closely related genera and species. Walking around a facility with a pheromone bait will often rouse otherwise hidden moths, especially if the premises is not otherwise baited.

During the day, adult moths rest, often on walls, pillars etc. and can often be seen. Unlike many moth species, pyralid moths do not respond well to light traps.

Larvae of these moths wander when mature and will often pupate in piles of sacking or corrugated cardboard, especially if undisturbed for a while.

Geographical distribution

Species	Pest status	USA & Canada	Central & South America	Europe & N.Asia	Mediterranean basin	Africa	S. & SE. Asia	Australia & Oceania
Cadra cautella	●●●●	X	X	X	X	X	X	X
Cadra figulilella	●	X			X	X		
Corcyra cephalonica	●●		X		X	X	X	X
Ephestia calidella	●				X	X		
Ephestia elutella	●●●	X	X	X	X	X		X
Ephestia kuehniella	●●●	X	X	X	X	X	X	X
Plodia interpunctella	●●●●	X	X	X	X	X	X	X
Pyralis farinalis	●	X	X	X	X	X	X	X

Pest status: ● minor to ●●●● major pest
X: recorded

Major pest species are cosmopolitan. *C. cautella* and *P. interpunctella* are most often found in warm-temperate to tropical regions. *C. cephalonica* is restricted to tropical and sub-tropical regions. *E. kuehniella* and *E. elutella* are most prevalent in temperate regions. *E. calidella* and *E. figulilella* are largely restricted to areas with warm-temperate and Mediterranean climates. *P. farinalis* is widespread in temperate regions.

References

Aitken (1984), Cox and Bell (1991), Haines (1974, 1981, 1991), Ferguson (1987), Mound (1989), Solis (1999) and Weismann (1987).

Tineid moths
(Family: Tineidae)

Nemapogon granella	European grain moth
Tinea spp.	Case-bearing clothes moths
Tineola bisselliella	Clothes moth

Summary

Feeding strategies	secondary pest, scavenger
Commodities attacked	damp grain and grain products, clothes, carpets, upholstery made from animal fibres, occasionally dried meat
Distribution	mainly temperate regions
Economic importance	low on stored food, high on textiles and woollen goods
Eggs	laid on cracks and crevices
Larvae	active, external feeders
Adults	short lived, do not feed on commodity, can fly

Introduction

Larvae of moths of the family Tineidae feed wholly or mostly on dried material of animal origin. As a result, many are associated with the nests of rodents and birds where they feed on dried remains, hair and feathers. *Tinea* spp. and *Tineola bisselliella* are pests of products containing natural fibres such as wool. Another species, *Nemapogon granella*, is found in association with dried material of vegetable origin and is a minor pest of stored grain.

Identification

The heads of tineid moths are covered in rough erect scales which give a hairy appearance (Figure 207). Labial palps are short and not curved upwards. When fresh, fore wings of *N. granella* are about 7 mm long and are distinctively marked with dark brown/black blotches on a lighter background (Figure 208). In nature, adults of *Tinea* spp. and *T. bisselliella* are rarely seen and are undistinguished light to dark brown coloured moths (Figure 209). Fore wings of *Tinea* spp. are 4.5–8 mm long and are darker and/or patterned. Feeding damage caused by *Tinea* spp. or *T. bisselliella* can be used to identify infestations to genus, (see 'Type of damage and symptoms', below).

The classification of *Tinea* spp. has been subject to change. The species referred to as *Tinea pellionella* in literature prior to the late 1970s is now known to be a complex of 11 species, of which at least five are pests or potential pests of stored products (Robinson 1979, Robinson and Nielsen 1993).

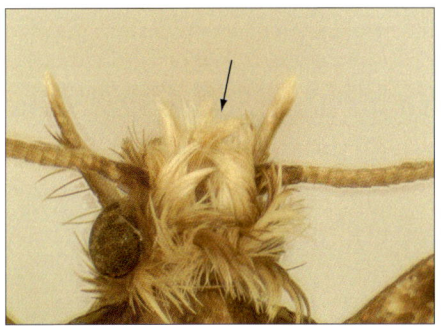

5 mm

5 mm

Top left: Figure 207 *Nemapogon granella*, adult, detail of head covered in rough hairs

Top right: Figure 208 *Nemapogon granella*, adult, pattern of fore wing, head covered in rough hairs

Left: Figure 209 *Tinea pellionella*, adult

Life cycle

Adults lay eggs in crevices. Larvae move over and through food material, feeding as they go. Pupation may occur in or near affected food material. Larvae of *N. granella* and *T. bisselliella* produce lots of webbing. *Tinea* spp. produce a silken tube-like case in which they live and pupate.

Physical limits and optimum rate of multiplication

Species	Conditions within which breeding takes place	Shortest development period, with optimum conditions	Maximum monthly rate of increase
Tineola bisselliella	10–33°C	39 days at 25°C (on fishmeal)	

Generation time is highly variable and depends on both environmental conditions and the nutritional quality of the infested material. Development data quoted above is rapid due to the high nutritional quality of the food media. Populations breeding on carpets etc. are likely to complete one to four generations a year. In cold winters they may overwinter in the larval stage.

Economic importance

N. granella can be a pest of stored grain usually associated with damp residues. It has also been reported attacking the corks of wine bottles, especially those that have mould growing upon them. *T. pellionella* and *T. bisselliella* are pests of articles containing wool and other material of animal origin such as feathers and horsehair. They are capable of causing significant damage to clothes, carpets, tapestries and upholstery. Occasionally they may attack stored products such as dried meat. As domestic pests these insects are most likely to be found in damp areas.

Type of damage and symptoms

Larvae of *N. granella* produce silk webbing and frass as they feed and pupate. Irregular holes may be bitten into attacked material. Larvae of *Tinea* spp. construct a portable cylindrical case of silk and cut fibres which they carry around and do not leave (Figures 210 and 211). By contrast, larvae of *T. bisselliella* produce mats and tubes of silk which become contaminated with frass and cast skins; they are often seen naked as they do not construct a portable case (Figure 212).

Figure 210 *Tinea dubiella*, silken tube containing cast skin of pupae

Figure 211 *Tinea dubiella*, infestation of underside of carpet

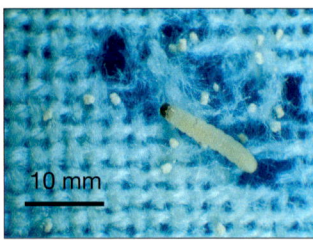
Figure 212 *Tineola bisselliella*, larvae and damage

Ecology

In nature tineid moths are frequent inhabitants of bird and animal nests and will also infest dried corpses of animals. The importance of these moths as domestic pests has declined due to increased use of artificial fibres and a general improvement in standards of building maintenance. Ecological requirements of the pest *Tinea* spp. vary. In Europe, for example, the installation of air conditioning and central heating is leading to the decline of *T. pellionella* and an increase of outbreaks of *T. dubiella*.

Geographical distribution

Species	Pest status	USA & Canada	Central & South America	Europe & N.Asia	Mediterranean basin	Africa	S. & SE. Asia	Australia & Oceania
Nemapogon granella	●	X		X	X			X
Tinea spp.	●●	X	X	X	X	X	X	X
Tineola bisselliella	●●	X	X	X	X	X	X	X

Pest status: ● minor to ●●●● major pest
X: recorded

N. granella is of temperate origin and is most often encountered in Europe and North America. *Tinea* spp. and *T. bisselliella* occur in mainly temperate and Mediterranean regions and in cool tropical areas. Infestations in other areas are mainly a result of importations.

References

Aitken (1984), Cox and Bell (1991), Mound (1989), Robinson (1979), Robinson and Nielsen (1993), Solis (1999) and Weismann (1987).

The order Psocoptera contains over 4000 species. Psocids are small soft-bodied insects that live on plant material, under bark, in leaf litter and in nests and animal burrows. Unlike beetles, moths and wasps, psocids undergo an incomplete metamorphosis where eggs hatch into nymphs, which are similar in appearance to adults except that they are smaller, paler and wingless. At each moult nymphs become bigger and more like the adult. Wings, where present become functional only in the adult stage.

Psocids, Booklice and Dustlice
(Families: Lachesillidae, Liposcelididae, Psyllipsocidae and Trogiidae)

Family: Lachesillidae

Lachesilla pedicularia

Lachesilla quercus

Family: Liposcelididae

Liposcelis bostrychophila

Liposcelis brunnea

Liposcelis corrodens

Liposcelis decolor

Liposcelis entomophila

Liposcelis paeta

Liposcelis pubescens

Liposcelis rufa

Family: Psyllipsocidae

Psocathorpos laclani

Family: Trogiidae

Lepinotus reticulatus

Lepinotus inqilinus

Lepinotus patruelis

Trogium pulsatorium

Summary

Feeding strategies	secondary pest, mould feeder
Commodities attacked	dried material of animal and especially plant origin
Distribution	worldwide
Economic importance	low to high
Eggs	laid in amongst commodity
Nymphs	similar in appearance to adults but smaller, mobile, external feeders
Adults	some are long lived, feed on commodity, Liposcelis wingless, others winged

Introduction

Species that are associated with stored products mostly belong to four families: the Lachesiilidae, Liposcelididae, Psyllipsocidae and Trogiidae (Figures 213–220). By far the most important of these is the Liposcelididae, which contains the wingless genus *Liposcelis*.

Identification

Genera of psocids associated with stored products can be identified as below. Identification of psocids to species is difficult and generally involves examination of cleared, prepared slide-mounted specimens at high magnification (see references below). Male specimens of *Liposcelis* spp. are mostly impossible to identify to species, except by association with females.

On account of their small size, wingless psocids can be confused with some larger predatory mites that occur in stored products. Mites have four pairs of legs, do not have antennae or wings and their bodies are not clearly divided into head, thorax and abdomen like insects. Predatory mites can be orange or red, colours never seen in psocids associated with stored products. *Liposcelis* species move with a characteristic rapid jerky motion, not seen in mite species.

Keys for the identification to species of psocids found in stored products can be found in Lienhard (1990, 1998) and Mockford (1987, 1993).

Figure 213 *Liposcelis bostrychophila*, adult showing long hair-like antennae and enlarged hind femur of hind leg typical of genus

Figure 214 *Liposcelis decolor*, adult, showing plain unmarked abdomen

Figure 215 *Liposcelis decolor*, infestation on surface of bulk stored barley

Figure 216 *Liposcelis entomophila*, adult, showing striped abdomen

Figure 217 *Lachesilla quercus*, adult, long wings which stretch well beyond end of abdomen

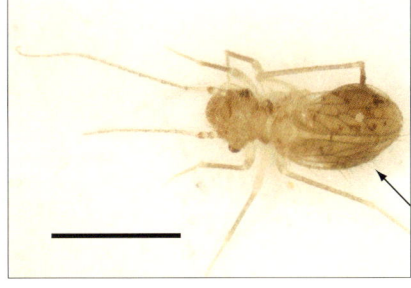

Figure 218 *Psocathorpos laclani*, adult, short wings which stretch to end of abdomen

Figure 219 *Lepinotus reticulatus*, adult, showing distinctive black colour and tiny scale-like wings

Figure 220 *Trogium pulsatorium*, adult, patterning on abdomen, dark line between eyes, tiny scale-like wings

Key to adults of major genera of Psocoptera associated with stored products

1 Body form highly flattened, adults wingless, length of adult – 0.7–1 mm, femur of hind leg much wider and flatter than other legs (Figure 214). Colour ranges from translucent to dark brown, (abdomen of adult *L. entomophila* characteristically striped – Figure 216) . . . . *Liposcelis* (Figures 213–216)
Body form globular – not flattened, adults winged which range from small scale like structures to full size wings with complete venation (Figures 217–220) 2

2 Wings as tiny scale like structures, fore wings only present (Figures 219– 220) 3
Wings larger, with venation, fore and hind wings present (Figures 217–218) 4

3 Colour of adult – frons (area of forehead between eyes) marked with a dark longitudinal line. Abdomen – light brown, with darker spots which form weak bands across abdomen; length of adult – 1.5–2.0 mm; adults can produce 'ticking' sound audible to humans . . . . *Trogium* (Figure 220)
Colour – adults black (nymphs pale grey, which get darker as they mature); length of adult – 1.2–1.5 mm . . . . *Lepinotus* (Figure 219)

4 Partly winged (brachypterous) – at rest tip of wings does not reach end of abdomen, length of adult 1.2–1.4 mm . . . . *Psocathorpos* (Figure 218)
Fully winged at rest tip of wings well beyond end of abdomen, flies readily; colour light brown, abdomen striped, venation of wings black, length of adult 2.0–2.7 mm . . . . *Lachesilla* (Figure 217)

Life cycle

The following data are for *Liposcelis* spp. Eggs are laid at random amongst the infested material. At 30°C, females lay about one egg per day which hatches in about a week. Metamorphosis is incomplete in Psocoptera and nymphs are similar in appearance to adults except they are smaller and paler. At 30°C development to adulthood takes about 21–28 days. Both adults and nymphs feed. Adults are long lived – several months at 30°C, six months or more at 20°C. Reproduction is sexual in all storage species, with the exception of *L. bostrychophila* which is parthenogenetic and consists only of females which lay fertile eggs without the intervention of males.

Physical limits and optimum conditions

Species	Conditions within which breeding takes place	Optimum conditions
Lachesilla quercus	16–30°C, r.h. > 70%	22–26°C, 70–80% r.h.
Liposcelis bostrychophila	20–36°C, r.h. > 60%	30°C, 80% r.h.
Liposcelis decolor	18–36°C, r.h. > 60%	32°C, 80% r.h.
Liposcelis entomophila	18–36°C, r.h. > 60%	30°C, 80% r.h.
Liposcelis paeta	24–42°C, r.h. > 60%	33°C, 70% r.h.

Liposcelis species breed most rapidly under warm humid conditions. *L. paeta* is exceptional by the standards of storage insects in general in its tolerance to temperatures above 40°C. Psocids are generally sensitive to low humidity. *Liposcelis*, for example, are unable to survive long term in locations where the mean relative humidity is below about 60% r.h. Temperatures required by *L. quercus* to breed are lower than for *Liposcelis* species – which reflects its temperate rather than tropical origin.

Economic importance

Liposcelis species are omnivorous and will eat almost any stored product of animal or plant origin. They also feed on moulds and on glues made from animal by-products.

Liposcelis are traditionally thought to be only minor pests of stored grain and grain products. Recent experience suggests this is not always so. Four species, *L. bostrychophila*, *L. decolor*, *L. entomophila*, and *L. paeta*, are by far the most important species as pests of stored products. Enormous populations have been reported, infesting grain stored in warm temperate and tropical regions. These cause at the very least severe contamination and possible rejection of the infested commodity. Worldwide, such outbreaks appear to be becoming more frequent. For example, during the 1990s in some parts of Australia, severe infestations of *Liposcelis* became much more frequent in bulk grain storage facilities.

Liposcelis spp., in particular *L. bostrychophila,* are also important pests in museums, libraries, food processing plants, retail and domestic premises. Newly constructed buildings can often become infested by a range of psocid species as they dry out. Infestations of psocids, especially *Liposcelis* spp., are increasingly being implicated as an important source of inhalation allergens. These may produce a range of respiratory problems in sensitised people.

Other genera listed are usually associated with damp material or locations and are generally minor pests, but may cause concern on account of their mere presence. Recently in Australia, *L. quercus* has become a significant nuisance pest in coastal grain handling facilities where large populations build up from time to time, causing distress and inconvenience to workers and contamination of storage structures.

Type of damage and symptoms

Liposcelis are secondary pests that are able to excavate the soft endosperm from damaged and broken grain. They feed preferentially on grain germ and are capable of completely eating out the germ by first gaining access to it via damage to the seed coat caused by harvesting and handling. Psocids will also eat mould spores and fungal hyphae. The small size and flattened form of *Liposcelis* spp. allows them easy access into all but the most well-sealed package. Psocids are a problem in food packing plants; in addition to attacking the product directly, they often become trapped under shrink wrapping used to secure boxes onto pallets. Their presence often leads to the rejection of the contaminated goods.

In museums and libraries psocids may attack books, paper and fabrics, especially if these items are slightly damp. They will feed on glues made from animal or plant material, especially if such material is hygroscopic.

In silos with severe infestations, accumulations of dead and dying *Liposcelis* may be found at the top of the cells (Figure 215) and in the chutes leading from the cell outlet. Similar accumulations also occur on nearby walkways where they can affect worker safety by making it very slippery underfoot.

Infestations of *Lachesilla* spp. produce quantities of silken webbing, which gets covered in dust etc.

Ecology

Psocids associated with grain storage can also be found in natural habitats, such as in leaf litter, bird nests, compost, under the bark of trees and in clumps of dry grass. Such populations provide a continuous source of reinfestation.

There appears to be a link between the incidence of heavy infestations of *Liposcelis* in grain stores and poor and inappropriate fumigation practice. Survival of the psocid may have been due to insufficient gas being used or rapid re-invasion after fumigation from the surrounding structure. Resulting populations of *Liposcelis* can be present in grain in the absence of other pests and predators. Populations freed from competition and predation may explode given suitable environmental conditions.

Liposcelis species are sensitive to mean relative humidities below 60%, roughly equivalent to a grain moisture content of about 13–14%. However, in Australia, heavy infestations of *Liposcelis* spp. are commonly encountered in dry grain of less than 11% moisture content. In buildings, psocid infestations are usually associated with dampness and poor ventilation.

Monitoring

Psocids are easy to trap. Pitfall traps inserted into grain bulks are effective. Corrugated cardboard makes a highly attractive crevice trap especially for *Liposcelis* species. Pieces of cardboard can be left on grain surfaces or on storage structures. Psocids which accumulate inside can simply be shaken out.

Geographical distribution

Species	Pest status	USA & Canada	Central & South America	Europe & N.Asia	Mediterranean basin	Africa	S. & SE. Asia	Australia & Oceania
Lachesilla pedicularia	••	X	X	X	X	X		X
Lachesilla quercus	••	?		X	X		X	X
Lepinotus reticulatus	••	X	X	X	X	X	X	X
Lepinotus inqilinus	••	X	X	X	X	X	X	X
Lepinotus patruelis	•	X	X	X	X	X		X
Liposcelis bostrychophila	•••	X	X	X	X	X	X	X
Liposcelis brunnea	••	X		X			X	X
Liposcelis corrodens	••	X	X	X	X		X	X
Liposcelis decolor	•••	X	X	X	X	X	X®	X
Liposcelis entomophila	•••	X	X	X	X	X	X	X
Liposcelis paeta	•••	X		X	X	X	X	X®
Liposcelis pubescens	••		X	X	X			X
Liposcelis rufa	•	X	X	X	X	X		X
Psocathorpos laclani	•	X	X		X	X	X	X
Trogium pulsatorium	•	X		X	X		X	X

Pest status: • minor to •••• major pest
X: recorded
®: restricted distribution
?: status unclear

Liposcelis spp. associated with stored products occur more or less worldwide in both temperate and tropical regions. Of those listed, *L. bostrichophila* is the most widely distributed species, with populations known from the tropics to remote sub-Antarctic islands. *L. entomophila* and *L. paeta* are often encountered in warm temperate to tropical climates. For example, in Australia, *L. entomophila* is the dominant species in northern grain-growing areas and is uncommon in southern states. In contrast, *L. decolor* dominates in southern Australia. Elsewhere, this species is a major pest in warm temperate and Mediterranean climates such as southern Europe and appears to be uncommon or absent in tropical regions.

Lepinotus spp., *Lachesilla* spp. and *T. pulsatorium* are most often encountered in temperate regions. Until recently, *L. quercus* was restricted to Europe and temperate Asia and has been intercepted in the USA. Recently, it has become established in Australia. *P. lachlani* appears to have a mostly tropical and sub-tropical distribution. In the USA, it is mostly restricted to southern states and in Australia is it restricted to the sub-tropical and tropical north-east.

References

Lienhard (1990, 1998), Lienhard and Smithers (2002), Mockford (1987, 1993), Rees (1994b).

The Hemiptera, or true bugs, have mouthparts that have been modified into a piercing needle like structure through which they suck liquid food. Some are plant feeders and are major pests such as aphids and leaf hoppers. Other are predators of insects and mites and some suck blood from humans and other mammals. Unlike beetles, moths and wasps, bugs undergo an incomplete metamorphosis: eggs hatch into nymphs that are similar in appearance to adults except that they are smaller and paler. At each moult, nymphs become bigger and more like the adult. Wings become functional only in the adult stage.

Predatory bugs
(Families: Anthocoridae and Reduviidae)

Flower bugs (Family: Anthocoridae)

Lyctocoris campestris	Stack bug
Xylocoris spp.	Cereal bugs

Assassin bugs (Family: Reduviidae)

Amphibolus venator
Peregrinator biannulipes

Summary

Feeding strategy	predator
Commodities attacked	other insects
Distribution	worldwide
Economic importance	beneficial insect
Eggs	laid amongst commodity
Nymphs	like small version of adults, without wings
Adults	feed, can fly

Introduction

Members of the Anthocoridae and Reduviidae are often found associated with stored products. They are predators of insects, such as psocids and the juvenile stages of moths and beetles. In nature, these species can also be found under bark, in galleries within fungi, in nests of birds and rodents and in decaying vegetable matter.

Identification

Anthocorid bugs are flattened, diamond- or triangular-shaped insects (Figures 221–222). Nymphs are pale in colour (*Xylocoris* nymphs are yellow or pink) and have wing buds or under-developed wings. Adults are brownish, often with lighter-coloured legs. When at rest, the tips of the fore wings are transparent and overlap over the middle of the abdomen. This overlap is seen as a pale diamond shape. Adult *Lyctocoris campestris* measure 3.5 to 4 mm long, *Xylocoris* a little smaller. Antennae are long and hair-like. The mouthparts consist of a long (as long as the antennae) straight needle-like structure which, when not in use, is held tucked neatly underneath the body between the bases of legs.

Reduviid bugs are bigger than anthocorids; adult *Peregrinator biannulipes* measure 6 to 7 mm long. Adults and nymphs are brown in colour. Mouthparts are strongly curved downwards and more heavily built than those of the anthocorid bugs. When not in use, mouthparts are held loosely, but not flush, under the head and thorax (Figure 223).

Top left: Figure 221 *Xylocoris* spp., adult, showing wing, basal half leathery and tip membranous

Top right: Figure 222 *Lyctocoris campestris*, adult, showing wings, basal half leathery and tip membranous

Left: Figure 223 *Peregrinator biannulipes*, adult, showing characteristic needle-like mouth part held curved between front legs when not in use

Life cycle

Eggs are laid at random in the vicinity of a suitable food supply. *L. campestris* lay up to 300 eggs in a lifetime, at a rate of up to six a day. Metamorphosis is incomplete in Hemiptera. Nymphs are similar in appearance to adults except they are wingless, smaller and paler (in the case of *Xylocoris* species pink or yellow in colour). Both adults and nymphs feed, and the size of prey they can tackle gets bigger as they approach adulthood.

At 30°C, *L. campestris* needs about four weeks to develop from egg to adult. At 26°C and 64% r.h. on a diet of mites and young fly larvae, *Xylocoris galactinus* took 22 days to develop from egg to adult. This species could tolerate temperatures up to 42°C. *A. venator* breeds between 21–32°C, 40–70% r.h. and will survive at temperatures between 13–41°C. Development to adult takes 40 to 165 days.

Economic importance

Predatory bugs are predators of a wide range of insect and mite pests and their presence can have a significant impact on pest populations. Their presence in large numbers in a storage facility indicates established pest populations.

Type of damage and symptoms

Predatory bugs do not damage stored commodities, but their presence may be an issue of contamination. Workers handling commodities containing these insects can sometimes complain about being bitten, especially by reduviid bugs.

Ecology

These insects are general predators of any insect or mite they are able to subdue, in particular soft-bodied life stages. The larger reduviid beetles can successfully attack adult beetles. *L. campestris* will also feed on blood of warm-blooded animals. The khapra beetle, *Trogoderma granarium* (Col.: Dermestidae), appears to be a favoured food item for *A. venator*.

Monitoring

Anthocorid bugs are very active and are easily seen (especially the pink / yellow nymphs) in a commodity when present in numbers. Reduviid bugs are harder to detect as they are usually present in smaller numbers and are more cryptic in their behaviour. Bugs may be captured in crevice and pitfall traps.

Geographical distribution

Species	Pest status	USA & Canada	Central & South America	Europe & N.Asia	Mediterranean basin	Africa	S. & SE. Asia	Australia & Oceania
Lyctocoris campestris	P	X	X	X	X	X	X	X
Xylocoris spp.	P	X	X	X	X	X	X	X
Amphibolus venator	P					X	X	
Peregrinator biannulipes	P		X			X	X	X

P: Predator
X: recorded

Anthocorid bugs are found worldwide, from cool temperate to tropical regions. Reduviid bugs are mostly tropical.

References
Aitken (1984), Haines (1974, 1981, 1991).

The Hymenoptera is a very large order of insects which includes familiar insects such as ants, saw-flies, bees, and wasps. Many hymenoptera are complex social insects and live in large colonies at the centre of which is a fertile female or queen. Some species of wasps are parasitoids of the eggs and larvae of beetles and moths. It is these wasps that are specifically associated with stored products.

Parasitic wasps
(Order: Hymenoptera)

Family: Ichneumonidae

Family: Bethylidae

Family: Braconidae

Family: Pteromalidae

Family: Trichogrammatidae

Summary

Feeding strategy	parasitoid of beetles and moths
Distribution	worldwide
Economic importance	beneficial insect
Eggs	laid on host – juvenile stage of host
Larvae	live in or on host
Adults	active, do not feed, most species winged

Introduction

The Hymenoptera specifically associated with stored products are parasitoids which attack the juvenile stages of beetle and moth pests. Ants can also be found in stores as incidental scavengers.

Identification

Parasitoid Hymenoptera are small wasp-like insects. They range in size from relatively large Ichneumonidae to the tiny Trichogrammatidae. Most are winged but some members of the family Bethylidae are wingless. Venation on the wing may be quite complex (network of veins) or reduced (a single vein along the leading edge of the wing). The junction between the thorax and abdomen is often very constricted to form a waist or petiole. In some species this has become stretched and the waist has become little more than a slender tube connecting the bulk of the abdomen to the thorax. Major families associated with stored products can be identified as follows:

Families of parasitic wasps associated with stored products

Family: Ichneumonidae (Figure 224)
- Large – body length > 5 mm
- Winged – wing venation full
- 'Waist' between abdomen and thorax long, ovipositor (a fine needle-like appendage at the tip of the abdomen, used for egg laying) relatively long and exposed

Figure 224 *Venturia canescens*, adult, large size, wings with full venation, obvious waist between thorax and abdomen, long ovipositor held externally

Family: Braconidae (Figure 225)
- Medium – body length 3–4 mm
- Winged – wing venation full
- 'Waist' between abdomen and thorax relatively short, ovipositor relatively short and exposed

Figure 225 *Habrobracon (Bracon)* spp., adult, wings with full venation, ovipositor held externally

Family: Pteromalidae (also Chalcididae, Encyrtidae and Eupelmidae) (Figures 226–227)
- Small – body length 1.5–3 mm
- Winged – wing venation reduced
- Head held vertically with mouthparts pointing downwards, body dark, metallic in colour

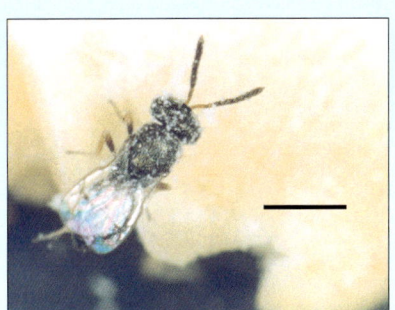

Figure 226 *Anisopteromalus calandrae*, adult, live

Figure 227 *Anisopteromalus calandrae*, adult, wing venation reduced, head held vertically with mouthparts pointing downwards

Family: Bethylidae (Figure 228)
- Small – body length 1.5–3 mm
- Winged or wingless – when winged wing venation reduced
- Head held horizontally with mouthparts pointing forwards, body dark, metallic in colour

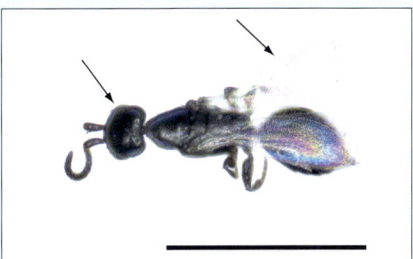

Figure 228 *Cephalonomia* spp., adult, wing venation reduced, head held horizontally with mouthparts pointing forwards

Trichogrammatidae
- Tiny – body length < 0.5 mm
- Winged – wing venation reduced

Life cycle

Detail of the life histories of parasitic wasps varies between families. Eggs are either laid into or onto the host. The larvae hatch and either live on the outside of their host or inside the host feeding on body tissues. At first wasp larvae are careful not to damage any vital organs which would cause the premature death of the host. The host is usually killed once larval development is completed or when the adult wasp emerges. The number of individual parasites able to survive in a host depends on the wasp species concerned and size of host. *Trichogramma* spp., for example, are so small that several wasps can complete development within a single moth egg. Adult wasps are generally short lived and do not feed.

Different wasp species are relatively host specific in the species of beetles and moths they attack (see below).

Host–parasite relationships of some parasitic wasps associated with stored products.

Family	Species	Life stage of host parasitised	Internal / external parasite	Hosts
Ichneumonidae	*Venturia canescens*	larvae	internal	Moths: Pyralidae – *Plodia*, *Cadra*, *Ephestia*, *Corcyra*
Braconidae	*Habrobracon (Bracon)* spp.	larvae	internal, emerges from larvae to pupate	Moths: Pyralidae – *Plodia*, *Cadra*, *Ephestia*, *Corcyra*
Pteromalidae	*Anisopteromalus calandrae*	larvae	internal	Beetles: especially *Sitophilus*, but also many other pest species Moths: *Sitotroga*
	Choetospila elegans	larvae	internal	Beetles: many pest species
	Dinarmus laticeps	larvae	internal	Beetles: *Callosobruchus*
	Lariophagus distinguendus	larvae	internal	Beetles: *Sitophilus*, *Rhyzopertha*, *Lasioderma*, *Stegobium*, Ptininae
Bethylidae	*Cephalonomia* spp.	larvae	external	Beetles: Anobiidae, *Sitophilus*, *Oryzaephilus*, *Cryptolestes*, Ptininae
	Holepyris hawaiiensis	larvae	external	Moths: *Plodia*, *Ephestia*
	Holepyris sylvanidis	larvae	external	Beetles: *Tribolium* spp.
	Laelius spp.	larvae	external	Beetles: Dermestidae
Trichogrammatidae	*Trichogramma* spp.	egg	internal	Moths: wide range of species
	Uscana spp.	egg	internal	Beetles: *Callosobruchus*

Physical limits and optimum rate of multiplication

The development time of a parasitic wasp is closely allied to the development rate of the host. Species that parasitise eggs complete their life cycle in a few days while those that attack larvae usually take longer, typically two to three weeks.

Economic importance

Parasitic wasps do not attack the stored product in which they are found. By attacking pest species, they perform a useful service and studies have shown that they can have a significant impact on the populations of many storage pests. However, for many stored products, especially processed products, the presence of any insect, even a 'beneficial one', is commercially unacceptable. There are situations, such as storage of raw products, subsistence agriculture and residual and structural infestation when encouragement or introduction of parasitic wasps as a control measure may be appropriate.

Type of damage and symptoms

Parasitic wasps do not damage stored commodities, but their presence may be an issue of contamination.

Ecology

The presence of large numbers of these insects usually indicates that a long-standing infestation is present. Large numbers of adult parasites can then be found walking over surfaces and flying near lights and windows searching for new hosts. Such populations are very easily killed by applications of pesticides such as surface sprays and fogging. Such treatments do little to control the pest within the bulk of a commodity and may result in the pest problem becoming worse as a result of, in effect, selective control of the parasites. Effective control would involve either complete elimination of all insects or possibly encouragement of parasitic or other beneficial insects.

Recently there has been more interest in using parasitic wasps to control pests under conditions of farm storage. Releases of wasps have also been undertaken to control residual populations of insects in factories and in empty storage structures.

Monitoring

Parasitic wasps are attracted to light and may often be found on window sills and around lights.

Geographical distribution

Parasitic wasps are found worldwide, from cool temperate areas to the tropics. The common species associated with stored products pests are found worldwide.

References

Haines (1991), Gordh (1987), Gordh and Hartman (1991).

Introduction

Storage pests are mostly small drab insects which are crepuscular in nature. Except when present in very high numbers, their presence is often not obvious to the untrained observer. Through a mixture of sampling, inspection and the use of traps an experienced store manager can detect low level infestations. When a population is detected early there is more time to plan and undertake effective pest control, before it gets out of hand and becomes obvious to everyone.

Sampling and inspection

Sampling of products

Very often grain and other products are sampled at intake, during transportation and use – mainly for assessment of grade and quality. Samples can also be examined for presence of insects. A variety of probes are used for either bulk or bagged grain (Figure 229). In many organisations, hand probes have been replaced with pneumatic sample probes which can be inserted into the back of a truck using a mechanical arm. Probe samples while adequate for quality assessment are not a very effective way of detecting insect infestation as the sample taken is typically very small relative to the load. Accumulations of insects in one part of the bulk are often missed by such sampling.

Grain moving through a handling system can be sampled using a diverter system. Here a small percentage of passing grain is continously diverted into a sampling system. Such systems have the advantage of delivering a representative sample for inspection. For example in Australian grain export facilities, such systems are set to deliver to the point of inspection a sample of 2.25 kg per

Figure 229 Sampling from top of silo using a hand-held probe

33 t of grain. To check for insects, grain is passed over a mechanical sieving machine which is attended by an inspector (Figure 230). For bagged commodities, the most reliable method of sampling is passing the contents of a number of randomly selected bags over an inclined sieve.

Figure 230 Mechanical sieve to extract insects from a stream of grain

Several companies have developed probes that can detect the noise that insects make when infesting a grain bulk. Some of these devices can compensate for ambient noise so can be used in industrial settings such as inspecting grain in the back of a truck. Other electronic sensors are being developed that claim to detect chemical odours released by insects – in the same way that hand-held devices are currently used to sniff for explosives.

Hidden infestations are often hard to detect. Radio photography using X-rays is effective. Other approaches including the development of PCR and ELISA tests and use of near-infrared spectroscopy and electronic 'nose' detectors are being researched.

Inspection of stored material and buildings

Insects are small and can be found almost anywhere within a building or storage or manufacturing facility. Given time, lack of disturbance and suitable temperature and humidity almost any edible residue of stored product will become infested. A quantity of a few grams can sustain an infestation.

It is critical that designers, engineers and fitters of grain and food handling equipment take care not to create places where infestable material may accumulate unnoticed. Very often, facilities with moderate insect problems look superficially clean and as a result management wonder why they have a problem. However, a careful search often leads to many foci of infestations being found. A torch, penknife and a paint brush are useful tools. From experience, such places invariably include:

- Dead spots in milling, handling and packing machines, elevator boots
- Rail cars, shipping containers, truck tippers, especially if made of wood

- Harvesting machinery, field bins, augers
- Rail/road receival grids
- On top, underneath and inside cupboards – especially those used by cleaners and fitters
- Floor sweepings (especially if stored up before being disposed of), brooms, vacuum cleaners
- Skips and dumpsters
- Locker rooms
- Kitchens and quality control laboratories
- Electrical switch boxes
- Cable trays, beams, pipes, conduits, shelving units
- Air conditioning ductwork
- Drains and traps
- Fire hose reels
- Gaps and conduits sealed with 'space invader' foam
- Expansion joints in walls and floors
- Desk draws, in and under computer keyboards
- Accumulations of paperwork, part-used boxes of tissues, disposable gloves etc.
- On top, behind and underneath just about anything, especially if it is not moved often
- Unused and decommissioned equipment, and office furniture, blanked off pipework
- Packaging material – such as boxes, sacks, bins, bulkerbags, pallets, especially if old and / or reused
- Sieving screens stored for repair and re-use
- Returned, damaged and out-of-date stock
- Infrequently used ingredients
- Rodent bait stations when baited with wax or granular bait
- Bird nests, dead rats and birds
- Long grass, reeds, under the bark of nearby trees
- Dumps, compost heaps.

Insects are usually more active in the evening and during warmer months. Some insects such as bruchids and the weevil *Sitophilus* may get agitated and leave a commodity that is shaken or dropped.

Trapping

The use of traps may assist inspectors in detecting and monitoring insect populations. Being in position for several days or weeks, traps often provide evidence of an infestation that could have occurred between inspection times. Many are highly attractive and can detect insects at level well below what is possible by simple inspection.

Traps fall into two types – those that rely on a physical feature to attract and detain insects and those that use a bait to attract insects – typically food or food extract or a synthetic sex or aggregation pheromone.

Insect species vary considerably in their propensity to be trapped and the type of trap that is most effective. The inspector needs to know which species they are targeting to select the right trap and bait. Species descriptions detail which traps are most useful for the particular group of insects. In grain bulks, major pests such as *Rhyzopertha dominica* and *Sitophilus* spp. are difficult to detect with traps whereas pests such as *Tribolium*, *Oryzaephilus*, *Cryptolestes* and *Liposcelis* species trap easily.

Traps are best used to detect the presence and geographical distribution of insects. For traps inserted into commodities, it is generally not possible to convert trap catch into number of insects per kg. Such relationships are highly situation specific as they are strongly affected by the insect species, commodity infested and the storage and environmental conditions. Very few such relationships have been worked out systematically and recorded in the scientific literature.

Before embarking on a trapping program you should be clear as to how the results obtained are to be used. For example, there is no point in using traps if the data is not going to be used to direct pest control operations. Equally, trapping may be pointless if insects are so abundant that they can easily be seen.

Types of traps

Crevice traps

Most pests that attack stored produce prefer to keep their body in contact with as much substrate as possible. Such insects will seek cracks and crevices rather than remain in the open. A trap which contains many attractive crevices will accumulate insects. Once insects enter they may be reluctant to leave even though they can.

A simple crevice trap can be made from a piece of corrugated cardboard (Figure 231). These traps are highly effective at detecting the psocid *Liposcelis* spp. and free-ranging beetles such as *Tribolium*, *Oryzaephilus* and *Cryptolestes*. The efficacy of such traps may be enhanced by a food bait or pheromone. Several commercial trapping systems take this approach. Crevice traps are especially useful for detecting insects in storage structures and empty storage bins.

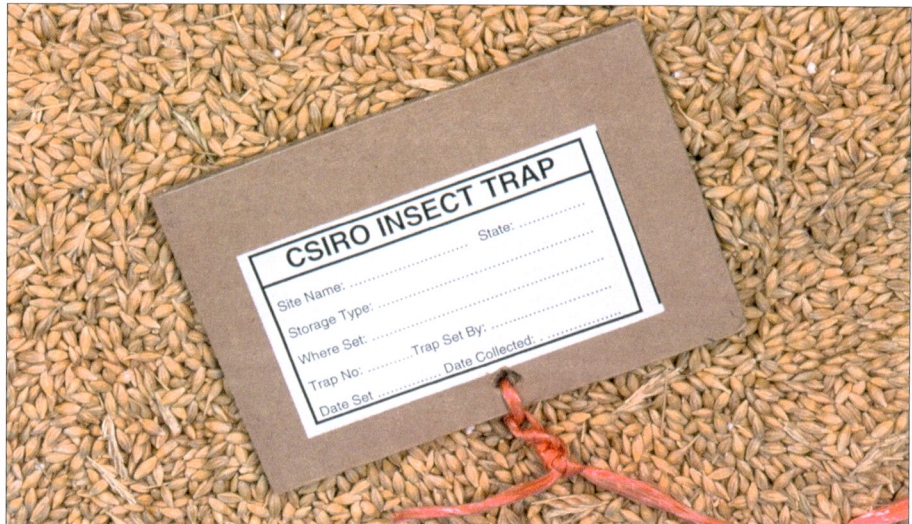

Figure 231 Crevice trap made from a piece of corrugated cardboard trap

Pitfall traps

Pitfall traps rely on the fact that many insects cannot climb out of a vertically walled container once they have fallen into it. Simple pitfall traps can be made from disposable plastic drinking cups or used drink cans (Figure 232). These can be buried into grain so the top is level with the grain surface. Grain can be prevented from entering the trap by covering the opening with mesh of a gauge that still allows insects to pass through. The sides of the trap may be coated with a PTFE suspension to make them slippery to those pests that would otherwise be able to climb out – notably *Oryzaephilus* spp. Alternatively, a small quantity of a food grade vegetable oil (e.g. cooking oil) may be placed in the bottom to trap insects that fall in.

A number of commercial trap designs based on this principle are available to catch insects on hard surfaces and for inserting deep with the grain (Figure 233). The effectiveness of these traps can be improved with the addition of pheromone or food bait. These traps are highly effective tools in the detection of free-ranging beetles such as *Tribolium*, *Oryzaephilus*, *Cryptolestes* and the psocid *Liposcelis* in or on grain bulks. They are less effective in detecting *Sitophilus* species and *Rhyzopertha dominica*. Researchers have been experimenting with pitfall traps that automatically count and even identify insects as they are captured and relay this information by radio or wire to the store manager.

Figure 232 Simple pitfall trap made from a disposable drink cup

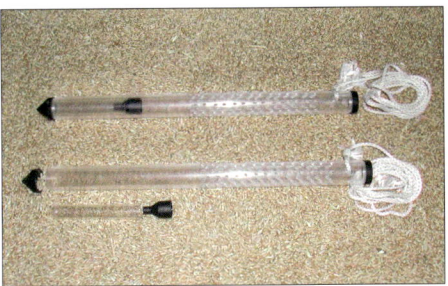 

Figure 233 Commercial pitfall traps for insertion into grain bulks

Bait traps

Certain foodstuffs are highly attractive to storage insects, especially those which contain volatile oils. Examples include dried carob, dried fruit, groundnuts and brown rice. A simple trap can be made from a mesh bag containing a mix of such materials (Figure 234). Insects attracted can simply be shaken out and examined. Bait traps work best in situations where there is little other attractive food about or the food in the trap is much more attractive than that in the store, e.g. brown rice when compared to milled rice.

Traps containing food can be lost and then they constitute a hazard as they will become a focus of infestation. A number of commercial trapping systems get round this problem by using a food oil based bait.

Figure 234 Bait bag made from plastic mesh

Light traps

Many insects are attracted to light, especially to the green to ultraviolet portion of the spectrum. Light traps have in the past been used to monitor presence of *Lasioderma serricorne* in the tobacco industry. Unfortunately the pyralid moths (*Cadra*, *Ephestia* and *Plodia*) are not very attracted to light. These are best detected using pheromone-based traps (see below).

Pheromone-baited traps

Figure 235 Pheromone-baited sticky flight trap

Pheromones are chemicals which insects release to attract a mate or other members of their species. Synthetic versions of pheromones of many storage insects are commercially available as baits. Baits are packaged in a vial or membrane that permits the slow release of the volatile chemical over weeks or months.

The synthetic sex pheromones of moths *Cadra*, *Ephestia* and *Plodia* are a highly effective bait for flight traps for these moths (Figure 235). Synthetic aggregation pheromones of *Lasioderma serricorne* / *Stegobium paniceum*, *Tribolium* spp., *Rhyzopertha dominica*, *Prostephanus truncatus* and *Trogoderma variabile* have also been extensively used in traps. They have proved especially useful in detecting outdoor populations of the last three species, which are almost impossible to detect by direct inspection.

Placement of traps

If using a commercial trapping system, the manufacturer or distributor should provide instructions for its use, in particular how many traps to use in a given area and how long any bait will remain effective. Avoid placing traps in high traffic areas where they stand a high chance of being destroyed or damaged by machinery or vehicles. Place or hang them out of harm's way, as insects are anyway more likely to be present in less visited areas. Ensure that workers and cleaning staff are aware of the traps and the need not to disturb them. Service traps on a regular basis as forgotten traps can sometimes be a source of infestation. Always label traps in a way that allows for accurate and easy data recording. When traps are used outside, any label should be weatherproof. If traps are deployed in locations to which the public has access, details of the traps' purpose together with contact details should be affixed to each trap.

When placing traps into commodities or close to product lines, care should be taken to eliminate or at least minimise risk of the trap or any of its components being lost or falling into the product.

Some suppliers and manufactures of pheromone lures and traps for stored product insects include

- Agrisense http://www.agrisense.demon.co.uk/
- Insects Limited http://www.insectslimited.com/
- Trécé http://www.trece.com/stgdprod.html

References

A selection of references on this subject:

Anon (1999), Arbogast et al. (2003), Atui et al. (2003), Bonjour and Phillips (2003), Burkholder (1984), Chambers (2003), Collins et al. (2003), Dowell et al. (1999), Gentry (1984), Gilbert (1984), Hodges et al. (2003), Hodges et al. (1985), Magan and Evans (2000), Mills and Pederson (1991), Mueller (1998), Rees (1999a), Rees (1999b), Shuman and Epsky (1999), Shuman et al. (2003), Subramanyam and Hagstrum (1996), Weier (2003), Wright (1991).

Collecting, preserving and shipping specimens for identification

Persons working in storage facilities may wish to build a collection of the insects they find. This can be useful for training purposes or for comparative identification of specimens. You will find insects not illustrated in this book and you may also want a second opinion about a particular specimen.

Collection data

It is most important that you collect basic data about each specimen; without this information, a specimen is greatly devalued. Our memories are imperfect and we all forget in time. Useful data to collect includes:

- **Place of collection** – e.g. state, town, location, name of farm.
- **Nature of location** – e.g. silo, shed, flour mill, cell number.
- **Commodity from which specimen was collected, how collected**
- **Date of collection**
- **Person who collected it** – name, phone number, email address.

Record this information at the time of collection on a small piece of paper. Put it into the collection tube or on a label firmly attached to the outside. Use a pencil or a permanent ink pen. If a pen is used, check the ink used is not soluble in alcohol; pencil is always safe. Do not use ballpoint pens, texta pens or similar as the inks used can be water or alcohol soluble. To avoid potential confusion it is best to write the month as words (as below) rather than as a number.

An example of a specimen data label:

> The black stump (State)
> XYZ flour millers Ltd.
> Wheat flour, bin B2
> 12 Sept 2003
> A. Smith, (00) 123 456

Other useful information (e.g. pesticide used, grain temperature/moisture content, time in store) may also be recorded.

Handling insect specimens

Insects are delicate and are easily damaged by rough handling. Avoid picking them up with your fingers. Use a wetted artist's paint brush, a scrap of paper or tip of a penknife blade to get them into a container. A torch is useful to find insects and see what you are doing.

Preservation of specimens

Insects can be preserved wet (e.g. in alcohol) or dry. Soft-bodied insects are best preserved wet.

Wet preservation

A solution of 70% alcohol (e.g. 7 parts ethanol or methylated spirits to 3 parts clean water) will kill and preserve specimens. Undiluted spirits (gin, vodka etc.) can be used for short-term preservation. Small glass or plastic tubes or bottles used must have air and liquid-tight screw-on or push-in tops. A wide range can be purchased from scientific suppliers. Do not use any container that leaks if shaken, squeezed gently or when turned upside down. See-through 35 mm film canisters make useful easily available field collection containers. Be aware that food processing enterprises often do not permit glass to enter their facilities. Only plastic tubes should be taken into such places.

Fill the tube with fluid to cover specimens. Insert a wad of paper tissue (not cotton wool) to hold insects down. Release trapped air bubbles with a pencil or unfolded paper clip. If specimen is to be transported by post or courier, secure the lid with masking or electricians tape. This is especially important if air transportation is envisaged as pressure changes experienced can make seemingly well sealed containers leak.

Dry preservation

Hard-bodied insects such as adult moths and beetles can also be preserved dry. They may be pinned or pointed for formal display in reference collections – see Upton (1991). For simple field preservation, insects may be put in a tube as above and held in place lightly with soft tissue paper. Specimens can be quickly killed by placing the tube in a domestic freezer. Dry preservation is especially useful if specimens are to be sent by air where restrictions may be in place concerning the carriage of inflammable liquids such as alcohol.

Getting help from a specialist

Sometimes there is a need for a specialist to confirm the identity of a specimen. Before sending specimens it is important that agreement is first obtained that the person in question, especially as these days it is common for institutions to charge for their services and these costs can be considerable. In seeking help you need to consider what level of effort is needed – do you want a quick visual confirmation or is the job likely to involve time-consuming dissection and specimen preparation. Any specimen provided should be in a clean and well prepared a state as is possible, few specialists are interested in examining large quantities of unsorted material. Specialists are busy people and are more likely to agree to examine one or two specimens rather than many, so only ask for assistance when is it really needed. If you want your specimens returned, you should make this known during initial negotiations.

Make sure you provide return and contact details. An email address will probably get the most rapid response as far as delivery of information is concerned.

A list of websites of organisations who may be able to offer assistance is given on page 168–169. In addition, local and national natural history museums and local, state or federal agriculture or plant health departments may also be able to offer assistance.

Packing and shipping

Care should be taken in packing specimens for transportation. Tubes are best sent in a box protected by suitable padding such as bubble wrap or clean shredded office paper. As a second best padded ('jiffy' type) bags can be used, but these offer less crush protection than a box. Wrap or pack tubes sufficiently to stop them moving about inside the box or bag. Never send specimens or tubes loose in ordinary envelopes, they are likely to be lost or damaged in transit. Crushed tubes etc. may also cause injury to postal workers.

Be aware of and follow any requirements by authorities to comply with postal legislations and rules concerning the movement of grain and insect specimens between states and overseas. This includes requirements in both the place of sending **and** the destination, especially if in another country.

References and resources

Printed material

The following list includes some of the most important and useful texts written on the biology and identification of stored product pests. Unfortunately many of these are out of print and/or available only in specialist libraries. This field guide is a review and summary of information presented in the references listed below.

Aitken, A.D. (1975) Insect Travellers. Volume 1. Coleoptera. *Technical Bulletin 31*, HSMO, xvi. Ministry of Agriculture Fisheries and Food: London, UK.

Aitken, A.D. (1984) Insect Travellers. Volume 2. *Reference Book 437*, HSMO, ix. Ministry of Agriculture Fisheries and Food: London, UK.

Anderson, D.M. (1987) Larval beetles (Coleoptera) In 'Insect and mite pests in food: An illustrated key'. (Ed. J.R. Gorham). *USDA Agriculture Handbook No. 655*. United States Department of Agriculture: Washington DC, USA. Vol. 1, pp. 95–136.

Anon (1993) Larger grain borer. *GASGA Technical Leaflet No. 1*. CTA: Wageningen, The Netherlands.

Anon (1999) *The Grain Storage Guide*. Home Grown Cereals Authority: London, UK.

Arbogast, R.T. (1991) Beetles: Coleoptera, In 'Ecology and management of food industry pests'. (Ed. J.R. Gorham). *FDA Bulletin 4*. Food and Drug Administration: Washington DC, USA. pp. 131–176.

Arbogast, R.T., Kendra, P.E., Chini S.R. and McGovern, J.E. (2003) Meaning and practical value of special analysis for protecting retail stores. In 'Advances in stored product protection', *Proceedings of the 8th International Working Conference on Stored Product Protection*. (Eds P.F. Credland, D.M. Armitage, C.H. Bell, P.M. Cogan, and E. Highley). York, UK, 22–26 July 2002, pp. 1033–1038.

Atui, M.B., Larazzari, S.M.N., Lazzari, F.A. and Flinn, P.W. (2003) Comparison of ELISA and fragment count methods for detection of insects in wheat flour. In 'Advances in stored product protection', *Proceedings of the 8th International Working Conference on Stored Product Protection*. (Eds P.F. Credland, D.M. Armitage, C.H. Bell, P.M. Cogan, and E. Highley). York, UK, 22–26 July 2002, pp. 135–137.

Banks, H.J. (1980) Identification of stored product *Cryptolestes* spp. (Coleoptera: Cucujidae) a rapid technique for the preparation of suitable mounts. *Journal of Australian Entomological Society*, **18**, 217–222.

Banks, H.J. (1994) Illustrated identification keys for *Trogoderma granarium*, *T. glabrum*, *T. inclusum* and *T. variable* (Coleoptera: Dermestidae) and other *Trogoderma* associated with stored products. *CSIRO Division of Entomology Technical Paper No. 32*, CSIRO, Canberra, Australia.

Bonjour, E.L. and Phillips, T.W. (2003) Comparing insect captures in the 'StorMax Insector' and other probe traps. In 'Advances in stored product protection', *Proceedings of the 8th International Working Conference on Stored Product Protection*. (Eds P.F. Credland, D.M. Armitage, C.H. Bell, P.M. Cogan, and E. Highley). York, UK, 22–26 July 2002, pp. 238–240.

Bousquet, Y. (1990) *Beetles Associated with Stored Products in Canada: An Identification Guide*. Research branch, Agriculture Canada: Ottawa, Canada.

Buckland, P.C. (1981) The early dispersal of insect pests of stored products as indicated by archaeological records. *Journal of Stored Products Research*, **17**, 1–12.

Burkholder, W.E. (1984) Use of pheromones and food attractants for monitoring and trapping stored-product insects. In *Insect Management for Food Storage and Processing*. (Ed. F.J. Baur). American Association of Cereal Chemists: St Paul, MN, USA. pp. 69–86.

Carvalho, Ed. Luna De (1979) Guia pratico para a idntificacao de alguns insectos de armazens e productos armazenados. Junta de Investigacoes cientificas de ultramar, Centro de defesa fitossanitaria dos produtos ultramarines. 3 volumes, Lisboa, Portugal.

Chambers, J. (2003) Where does pest detection research go next, Keynote paper. In 'Advances in stored product protection', *Proceedings of the 8th International Working Conference on Stored Product Protection*. (Eds P.F. Credland, D.M. Armitage, C.H. Bell, P.M. Cogan, and E. Highley). York, UK, 22–26 July 2002. pp. 103–109.

Collins, L.E. Chambers, J., and Cogan, P. (2003) The i-Spy Insect Indicator™: development of an insect monitoring trap for use on flat surfaces in the cereal and food trades, and potential applications. In 'Advances in stored product protection', *Proceedings of the 8th International Working Conference on Stored Product Protection*. (Eds P.F. Credland, D.M. Armitage, C.H. Bell, P.M. Cogan, and E. Highley). York, UK, 22–26 July 2002. pp. 196–199.

Connell, W.A (1987) Sap beetles, (Nitidulidae, Coleoptera). In 'Insect and mite pests in food: an illustrated key'. (Ed. J.R. Gorham). *USDA Agriculture Handbook No. 655*. United States Department of Agriculture: Washington DC, USA. Vol. 1, pp. 115–136.

Cox, P.D., and Bell C.H. (1991) Biology and ecology of moth pests of stored food. In 'Ecology and management of food industry pests'. (Ed. J.R. Gorham). *FDA Bulletin 4*. Food and Drug Administration: Washington DC, USA.

Dobson, R.M. (1952) The species of *Carpophilus* Stephens (Col. Nitidulidae) from Australia. *Entomologists Monthly Magazine*, **88**, 256–258.

Dobson, R.M. (1954a) The species of *Carpophilus* Stephens (Col. Nitidulidae) associated with stored products. *Entomologists Monthly Magazine*, **92**, 41–42.

Dobson, R.M. (1954b) The species of *Carpophilus* Stephens (Col. Nitidulidae) associated with stored products. *Bulletin of Entomological Research*, **92**, 389–402.

Dowell, F.E., Thorne, J.E. Wang, D., and Baker, J.E. (1999) Identifying stored-product insects using near-infrared spectroscopy, *Journal of Economic Entomology*, **92**, 165–169.

Ferguson, D.C. (1987) Adult moths (Lepidoptera) In 'Insect and mite pests in food: An illustrated key'. (Ed. J.R. Gorham). *USDA Agriculture Handbook No. 655*. United States Department of Agriculture: Washington DC, USA. Vol. 1, pp. 231–244.

Gentry, J.W. (1984) Inspection techniques. In 'Insect management for food storage and processing'. (Ed. F.J. Baur).American Association of Cereal Chemists: St Paul, MN, USA. pp. 33–42.

Gilbert, D. (1984) Insect electrocutor light traps In 'Insect management for food storage and processing'. (Ed. F.J. Baur). American Association of Cereal Chemists: St Paul, MN, USA. pp. 87–108.

Gordh, G. (1987) Parasitic wasps (Apocrita, Hymenoptera). In 'Insect and mite pests in food: An illustrated key'. (Ed. J.R. Gorham). *USDA Agriculture Handbook No. 655*. United States Department of Agriculture: Washington DC, USA. Vol. 2, pp. 449–477.

Gordh, G. and Hartman, H. (1991) Hymenopterus parasites of stored-food insect pests. In 'Ecology and management of food industry pests'. (Ed. J.R. Gorham). *FDA Bulletin 4*. Food and Drug Administration: Washington DC, USA.

Gorham, J.R. (Ed.) (1991) 'Ecology and management of food industry pests'. *FDA Bulletin 4*. Food and Drug Administration: Washington DC, USA.

Haines, C. P. (1974) Insect and arachnids from stored products: report on specimens received by the Tropical Stored Products Centre 1972–1973. *Report L39*, 22 pp. Tropical Products Institute: London, UK, (now, Natural Resources Institute, Chatham, Kent, UK).

Haines, C. P. (1981) Insect and arachnids from stored products: report on specimens received by the Tropical Stored Products Centre 1973–1977. *Report L54*, 73 pp. Tropical Products Institute: London, UK, (now, Natural Resources Institute, Chatham, Kent, UK).

Haines, C.P. (1989) Observations on *Callosobruchus analis* (F.) in Indonesia, Including a key to storage *Callosobruchus* spp. (Col., Bruchidae). *Journal of Stored Products Research*, 25, 9–16.

Haines, C.P. (Ed.) (1991) *Insect and Arachnids of Tropical Stored Products: Their Biology and Identification.* Natural Resources Institute: Chatham, Kent, UK.

Haines, C.P. and Rees D.P. (1989) A field guide to the types of insects and mites infesting cured fish. *FAO Fisheries Technical paper No. 303.* Food and Agriculture Organization, Rome.

Halstead, D.G.H. (1986) Keys for the identification of beetles associated with stored products. I – Introduction and keys to families. *Journal of Stored Products Research*, 22, 163–203.

Halstead, D.G.H. (1993) Keys for the identification of beetles associated with stored products. II–Laemophloeidae, Passandridae and Silvanidae. *Journal of Stored Products Research*, 29, 99–197.

Harney, M. (1993) A guide to the insects of stored grain in South Africa. *Plant Protection Research Institute Handbook No. 1.* Agricultural Research Council: Pretoria, South Africa.

Hinton, H.E. (1940) The Ptinidae of economic importance. *Bulletin of Entomological Research*, 31, 331–381.

Hinton, H.E. (1941) The Lathridiidae of economic importance. *Bulletin of Entomological Research*, 32, 191–247.

Hinton, H.E. (1945a) The Histeridae associated with stored products. *Bulletin of Entomological Research*, 35, 309–340.

Hinton, H.E. (1945b) A *Monograph of the Beetles Associated with Stored Products.* Volume 1. British Museum (Natural History), London, UK.

Hodges, R.J. (1986) The biology and control of *Prostephanus truncatus* – a destructive pest with an increasing range. *Journal of Stored Products Research*, 22, 1–14.

Hodges, R.J. (1994) Recent advances in the biology and control of *Prostephanus truncatus* – (Coleoptera: Bostrichidae). In 'Stored product protection', *Proceedings of the 6th International Working Conference on Stored Product Protection.* (Eds E. Highley, E.J. Wright, H.J. Banks and B.R. Champ). Canberra, Australia, 17–23 April 1994. pp. 929–934.

Hodges, R.J., Birkenshaw, L.A. and Addo, S. (2003) Warning farmers when risk of infestation by *Prostephanus truncatus* is high. In 'Advances in stored product protection', *Proceedings of the 8th International Working Conference on Stored Product Protection.* (Eds P.F. Credland, D.M. Armitage, C.H. Bell, P.M. Cogan, and E. Highley). York, UK, 22–26 July 2002. pp. 110–114.

Hodges, R.J., Halid, H., Rees, D.P. Meik, J. and Sarjono, J. (1985) Insect traps tested as an aid to pest management in milled rice stores. *Journal of Stored Product Research*, 21, 215–219.

Howe, R.W. (1965) A summary of the estimates of optimal and minimal conditions for population increase of some stored products insects. *Journal of Stored Products Research*, 1, 177–184.

Howe, R.W. (1991) Spider beetles: Ptinidae. In: 'Ecology and management of food industry pests'. (Ed. J.R. Gorham). *FDA Bulletin 4.* Food and Drug Administration: Washington DC, USA.

Hughes, A.M. (1976) The mites of stored food and houses. Ministry of Agriculture, Fisheries and Food, *Technical Bulletin No. 9.* (Second edition). HMSO, London.

Kingsolver J.M. (1987) Dermestid beetles (Dermestid beetles Dermestidae, Coleoptera) In 'Insect and mite pests in food: an illustrated key'. (Ed. J.R. Gorham). *USDA Handbook No. 655*, United States Department of Agriculture: Washington D.C. USA. Vol. 1, pp. 115–136.

Kingsolver, J.M. and Andrews, F.G. (1987) Minute mould beetles (Lathridiidae, Coleoptera) In 'Insect and mite pests in food: an illustrated key'. (Ed. J.R. Gorham). *USDA Handbook No. 655*, United States Department of Agriculture: Washington DC, USA. Vol. 1, pp. 179–184.

Lienhard, C. (1990) Revision of the Western Palaearctic species of *Liposcelis* Motschulsky (Psocoptera: Liposcelidae). *Zoologishe Jahrbuecher Abteilung fur Systematik Oekologie und Geographie der Tiere*, 117, 117–174.

Lienhard, C. (1998) Psocopteres Euro-mediterraneens, Faune de France, 83, Federation Francaise des Societes de Sciences Naturrelles.

Lienhard, C. and Smithers, C.N. (2002) Psocoptera (Insecta). *World Catalogue and Bibliography.* Instrumenta Biodiversitatis V, Natural History Museum, Genève, Switzerland.

Magan, N. and Evans, P. (2000) Volatiles in grain as an indicator of fungal spoilage, odour descriptors for classifying spoiled grain and the potential for early detection using electronic nose technology: A review. *Journal of Stored Product Protection*, 36, 319–340.

Mills, R. and Pederson J. (1991) *A Flour Mill Sanitation Manual.* Eagan Press: St Paul, MN, USA.

Mockford, E.L. (1987) Psocids (Psocoptera). In 'Insect and mite pests in food: an illustrated key'. (Ed. J.R. Gorham). *USDA Agriculture Handbook No. 655.* United States Department of Agriculture: Washington DC, USA. Vol. 2, pp. 371–402.

Mockford, E.L. (1993) 'North American Psocoptera', *Flora and Fauna Handbook No. 10.* Sandhill Crane Press Inc: Gainesville, Florida, USA.

Mound, L. (1989) *Common Insect Pest of Stored Food Products.* 7th edn. Economic Series No. 15. British Museum (Natural History): London, UK.

Mueller, D.K. (1998) *Stored Product Protection. A Period of Transition.* Insects Limited: Indianapolis, USA.

Peacock, E.R. (1993) Adults and larvae of hide, larder and carpet beetles and their relatives (Coleoptera: Dermestidae) and of derodontid beetles (Coleoptera: Derodontidae). *Handbooks for the Identification of British Insects, Vol. 5, Part 3.* Royal Entomological Society of London: London, UK.

Prakash, A., Rao, J., Jayas, D.S., and Allotey, J. (Eds) (2003) *Insect Pests of Stored Products: A Global Scenario.* Applied Zoologists Research Association: Central Rice Research Institute, Cuttack, India.

Rees, D.P. (1994) Studies of the distribution and control of Pscoptera (psocids or booklice) associated with the grain industry in Australia. *CSIRO Division of Entomology report No. 57.* CSIRO: Canberra, Australia.

Rees, D. (1999a) Do-it-yourself traps to monitor storage pests, *Australian Grain*, 9, 3–4.

Rees. D. (1999b) Estimation of the optimum number of pheromone baited flight traps needed to monitor phycitine moths (*Ephestia cautella* and *Plodia interpunctella*) at a breakfast cereal factory – a case study. In 'Stored product protection', *Proceedings of the 7th International Working Conference on Stored Product Protection.* (Eds J. Zuxun, L. Quan, L. Yongsheng, T. Xianchang and G. Lianhua). Beijing, P.R. China, 14–19 October 1998. Vol. 2, pp. 1464–1471.

Roach, A.M.E. (2000) Review of the Australian species of the dermestid genus *Anthrenocerus* Arrow (Coleoptera : Dermestidae), *Invertebrate Taxonomy*, 14, 175–224.

Robinson, G.S (1979) Cloths moths of the *Tinea pellionella* complex: a revision of the world's species (Lepidoptera Tineidae), *Bulletin of the British Museum (Natural History)*, 38, 57–128.

Robinson, G.S and Nielsen, E.S. (1993) Tineid Genera of Australia (Lepidoptera), *Monographs on Australian Lepidoptera*, Volume 2, CSIRO Publications, Melbourne, Australia.

Semple, R.L. (1985) Pest control in grain storage systems in the ASEAN Region, ASEAN crops post-harvest programme, *Technical Paper Series No. 1.* Manila, Philippines.

Shuman, D. and Epsky, N.D. (1999) Computerized monitoring of stored-product insect population. In 'Stored product protection', *Proceedings of the 7th International Working Conference on Stored Product Protection.* (Eds J. Zuxun, L. Quan, L. Yongsheng, T. Xianchang and G. Lianhua). Beijing, P.R. China, 14–19 October 1998. Vol. 2, pp. 1429–1436.

Shuman, D., Epsky, N.D. and Crompton, D. (2003) Commercialisation of a species identifying automated stored product insect monitoring system. In 'Advances in stored product protection', *Proceedings of the 8th International Working Conference on Stored Product Protection.* (Eds P.F.

Credland, D.M. Armitage, C.H. Bell, P.M. Cogan, and E. Highley). York, UK, 22–26 July 2002. pp. 144–150.

Southgate, B.J. (1979) Biology of the Bruchidae. *Annual Review of Entomology*, 24, 449–473.

Spilman, T.J. (1982) False powderpost beetles of the Genus *Dinoderus* in North America (Coleoptera, Bostrichidae), *Coleopterists Bulletin*, 36, 193-196.

Spilman T.J. (1987a) Spider beetles (Ptinidae, Coleoptera) In 'Insect and mite pests in food: an illustrated key'. (Ed. J.R. Gorham). *USDA Agriculture Handbook No. 655*. United States Department of Agriculture: Washington DC, USA. Vol. 1, pp. 137–148.

Spilmann, T.J. (1987b) Darkling beetles (Tenebrionidae, Coleoptera) In. 'Insect and mite pests in food: an illustrated key'. (Ed. J.R. Gorham). *USDA Handbook no 655*, Washington DC, USA. Vol. 1, pp. 185–214.

Sokoloff, A (1972-4) *The Biology of Tribolium with Special Emphasis on Genetic Aspects*, Volumes 1–3. Clarendon Press: Oxford.

Solis, M. A (1999) Key to selected Pyraloidea (Lepidoptera) Larvae intercepted at US ports of entry: Revision of Pyraloidea in 'Keys to some frequently intercepted lepidopterus larvae' by D.M. Weisman 1986, *Proceedings of the Entomological Society of Washington*, 101, 645–686.

Subramanyam, B. and Hagstrum, D.W. (1996) Sampling. In *Integrated Management of Insects*. (Eds B. Subramanyam and D.W. Hagstrum). Marcel Dekker: New York, USA. pp. 135–193.

Upton, M.S. (1991) Methods for collecting, preserving and studying insects and allied forms. 4th edn. *Miscellaneous Publication No. 3*, 86 pp. Australian Entomological Society: Brisbane, Australia.

Wang Suya, Shan Meijing (Eds.) (1998) Grain storage in China. *7th International Working Conference on Stored Product Protection*. Beijing, P.R. China, 14–19 October 1998.

Weismann D.M. (1987) Larval moths (Lepidoptera) In 'Insect and mite pests in food: an illustrated key'. (Ed. J.R. Gorham). *USDA Agriculture Handbook No. 655*. United States Department of Agriculture: Washington DC, USA. Vol. 1, pp. 245–268.

White, N.D.G. (1995) Insects, Mites and Insecticides in stored-grain ecosystems. In *Stored Grain Ecosystems*. (Eds. D.S. Jayas, N.D.G. White and W.E. Muir). Marcel Dekker: New York, USA. pp. 123–167.

Weier, J. A. (2003) Value of spatial analysis in pest management from the perspective of a pest control operator. In 'Advances in stored product protection', *Proceedings of the 8th International Working Conference on Stored Product Protection*. (Eds P.F. Credland, D.M. Armitage, C.H. Bell, P.M. Cogan, and E. Highley), York, UK, 22–26 July 2002, pp. 1028–1032.

Wright, E.J. (1991) A trapping method to evaluate efficacy of a structural treatment in empty silos In: *Proceedings of the Fifth International Working Conference on Stored-product Protection*. (Eds F. Fleurat-Lessard and P. Ducom), Bordeaux, France, 9–14 September 1990, Vol. 3, pp. 1455–1463.

Zakladnoi, G. A. and Ratanova, V. F. (1987) Stored grain pests and their control. *Russian Translations Series 54*, A.A. Balkema: Rotterdam, The Netherlands.

The worldwide web

Most education and research facilities and Government agencies with an interest in stored product protection have a presence on the worldwide web. An alphabetic list is given below of sites which have some English content. This list below is not exhaustive and inclusion or otherwise on this list does not imply endorsement or otherwise by the author or CSIRO. Web addresses listed below are correct as of 8 June 2004. Websites have the habit of being moved, so over time these links may no longer work. Should this be the case, typing the name of the organisation into a search engine such as Google (www.google.com) will usually find a current way to enter the website. Another hazard is that organisations have their names changed from time to time or are even disbanded; in addition, new ones are started.

A selection of education and research facilities and Government agencies

- Agricultural Research Organization of Israel
  http://www.agri.gov.il/Volcani.html
- Australian Centre for International Agriculture Research
  http://www.aciar.gov.au
- Biologische Bundesanstalt für Land- und Forstwirtschaft, Germany
  http://www.bba.de/english/bbaeng.htm (in English)
  http://www.bba.de (in German)
- Canadian Grain Commission
  http://www.grainscanada.gc.ca/main-e.htm (English site)
- Canadian Grain Storage website hosted by Agriculture and AgriFood Canada, University of
  Manitoba and the Canadian Grain Commission
  http://res2.agr.ca/winnipeg/cgs_e.htm
- Central Food Technological Research Institute, India, Department of Food Protectants and
  Infestation Control
  http://www.cftri.com/department/foodpro.htm
- Central Science Laboratory, UK
  http://www.csl.gov.uk/main.cfm
- Danish Pest Infestation Laboratory
  http://www.dpil.dk/engindex.htm
- Department of Agriculture Western Australia, Entomology Branch
  http://www.agric.wa.gov.au/ento/index.htm
- Deutsche Gesellschaft für Technische Zusammenarbeit (GTZ) GmbH
  http://www.gtz.de/unternehmen/english/ (in English)
  http://www.gtz.de (in German)
- Food and Agriculture Organization of the United Nations, Information Network on Post-
  Harvest Operations
  http://www.fao.org/inpho
- Grains Research and Development Corporation, Australia
  http://www.grdc.com.au
- Institut National de la Recherche Agronomique, France
  http://www.inra.fr/ENG/ (in English)
  http://www.inra.fr (in French)
- International Institute of Tropical Agriculture
  http://www.iita.org
- International Maize and Wheat Improvement Center
  http://www.cimmyt.org
- Journal of Stored Products Research
  http://www.elsevier.com/locate/jspr
- Kansas State University (KSU), Department of Entomology
  http://www.oznet.ksu.edu/entomology/research_programs.htm#listings
- Natural Resources Institute, UK
  http://www.nri.org
- New South Wales Agriculture, Wagga Wagga Agriculture Research Institute, Australia
  http://www.agric.nsw.gov.au/reader/1145
- Proceedings of the International Working Conference on Stored-Product Protection
  (Tables of Contents)
  http://bru.gmprc.ksu.edu/proj/iwcspp/
- Purdue University, Lafayette, USA, Post Harvest Grain Quality and Stored Product
  Protection Program
  http://pasture.ecn.purdue.edu/~grainlab/

- Queensland Department of Primary Industries, Australia, research into stored product protection
  http://www.dpi.qld.gov.au/fieldcrops/3936.html
- Stored Grain Research Laboratory, CSIRO Entomology, Canberra, Australia
  http://www.sgrl.csiro.au
- United States Department of Agriculture
  http://www.usda.gov

Specific information on insect species

Typing the name of any insect into a search engine such as Google will usually produce lots of possible links. This is especially so for major pest species or storage insects which are popular as subjects for general biological research, such as *Tribolium* species.

Use of information on the internet

The standard and accuracy of the information available on the internet varies considerably. The sites listed above are likely to have subjected their information to a level of editorial checking before making it available to the public. Regardless of source, care needs also to be taken to ensure that information and advice are relevant to your location. Specific advice on the management of a pest, say in the USA, may or may not be directly relevant to a reader in Australia or tropical Africa as climatic, social and economic circumstances all differ. Particular care needs to be taken with recommendations on use of pesticides. What is legal in the country of recommendation may not be legal in the country where the information is accessed. It is always best to access information on pesticide use from your own responsible national authority.

Index to species

Insects covered by this book are listed by scientific name and by common name.

By scientific name

By common name

Figure credits

Figures 1, 3, 36, 41, 59, 73, 100, 103, 111, 116, 122, 128, 136, 146–147, 153, 159, 171, 176, 209–211, 214, 217, 221 and 226 by John Green, CSIRO

Figures 2, 4–5, 7–11, 13–26, 28–31, 33–35, 37, 39–40, 42–43, 45–58, 60, 62–65, 67–72, 75–97, 99, 101–102, 105, 107–110, 112–115, 117–121, 123–127, 129–135, 137–140, 143–145, 148–152, 154–158, 160–170, 172–174, 177–206, 208, 222–225 and 227–228 by Vanna Rangsi, CSIRO

Figures 6, 32, 38, 44, 61, 74, 98, 175, 207, 215, 218, 220 and 230 by David Rees, CSIRO

Figures 12, 213, 216, 219, 231–232 and 234 by David McClenaghan, CSIRO

Figures 27 and 106 by Natural Resources Institute, UK

Figures 66 and 104 by CSIRO

Figures 141, 142, 229, 233 and 235 by Noel Starick, CSIRO

Figure 212 by Roslyn Schuhmacher, CSIRO